Oracle 18c

必须掌握的新特性

管理与实战

戴明明　臧强磊　编著

電子工業出版社
Publishing House of Electronics Industry
北京•BEIJING

内 容 简 介

在 Oracle 12c 版本之前，数据库版本的迭代中基本架构都没有非常明显的变化，但从 Oracle 12c 版本开始，引入了很多新特性，其主要的特性颠覆了之前的概念，尤其是引入了多租户的概念。本书主要针对 Oracle 12c 版本以后的新特性进行讲解。本书在编写的过程中，以 Oracle 18c 为环境进行测试，内容涵盖 Oracle 18c 的多租户架构、In-Memory 特性的使用、ASMFD、Flex ASM，以及 RAC 集群环境中的 GIMR 和 CHM。通过学习本书的内容，读者可以快速掌握 Oracle 12c 和 Oracle 18c 的管理和使用方法。

本书不是一本基础的入门图书，在阅读本书时需要对 Oracle 的基本概念有一定的了解，同时具备一定的实际操作能力，本书适合 Oracle 运维人员和具备一定 Oracle 基础的开发人员阅读。

图书在版编目（CIP）数据

Oracle 18c必须掌握的新特性：管理与实战 / 戴明明，臧强磊编著. —北京：电子工业出版社，2019.7
ISBN 978-7-121-35502-8

Ⅰ. ①O… Ⅱ. ①戴… ②臧… Ⅲ. ①关系数据库系统 Ⅳ. ①TP311.138

中国版本图书馆CIP数据核字(2018)第258032号

责任编辑：戴　新
印　　刷：三河市良远印务有限公司
装　　订：三河市良远印务有限公司
出版发行：电子工业出版社
　　　　　北京市海淀区万寿路 173 信箱　　邮编：100036
开　　本：787×980　1/16　印张：20.5　字数：361 千字
版　　次：2019 年 7 月第 1 版
印　　次：2019 年 7 月第 1 次印刷
定　　价：79.00 元

凡所购买电子工业出版社图书有缺损问题，请向购买书店调换。若书店售缺，请与本社发行部联系，联系及邮购电话：（010）88254888，88258888。

质量投诉请发邮件至 zlts@phei.com.cn，盗版侵权举报请发邮件至 dbqq@phei.com.cn。

本书咨询联系方式：010-51260888-819，faq@phei.com.cn。

序 1

在 2014 年，编辑安娜约我写一本书，我欣然同意，但因为工作的原因，写作断断续续，一直没有定稿，后来 Oracle 版本又进行了迭代，就放弃了原计划。2016 年，安娜说戴老师，要不我们先翻译一本书吧，我又答应了，这一次安娜盯得比较紧，于是在 2018 年 1 月 1 日，我参与翻译的《Oracle 数据库问题解决方案和故障排除手册》一书上市。 因为写了很多年的博客，我也希望把自己的学习心得和方法记录下来，所以就有了本书。Oracle 从 12c 开始变化很大，不像从 10g 到 11g 的迭代，从 11g 到 12c 的迭代引入了很多新特性、新架构，这些新特性和新架构与之前版本的特性和架构完全不同，即使是我们这些使用 Oracle 多年的用户，也需要重新学习。在 2018 年 7 月 24 日，Oracle 18.3（18c）被正式发布，从版本上看，Oracle 18.3 相当于 12.2.0.2 版本。从以往的经验来看，Oracle 18c 是相对稳定的版本，加上 Oracle 11g 的生命周期也已经结束，所以未来几年，将会是 Oracle 18c（12.2.0.2）、19c（12.2.0.3）大规模使用的几年，DBA 们需要尽快掌握 Oracle 18c，而本书就是一本实用的辅助用书。

本书重点介绍 Oracle 18c 中最重要的几个新特性，如 Oracle 多租户架构管理、In-Memory 特性的使用、ASMFD、Flex ASM，以及 RAC 集群环境中的 GIMR 和 CHM。 这些都是有别于之前版本的特性，除这些特性外，其他的管理特性和之前数据库版本中的管理特性基本没有区别，所以读者在阅读本书之前，最好有一定的数据库基础，了解数据库的常规管理和操作，那么阅读起来就会比较顺畅。

我在翻译完《Oracle 数据库问题解决方案和故障排除手册》一书后和朋友说，翻译图书很累，主要是花费的时间远比预期的时间多。而写书相对轻松一点，可以按照自己的想法来编排章节的内容，但工作量也不少，先要梳理概念，再进行测试，最后进行整理。有些概念不太好理解，需要反复进行分类测试，最终才能确定。在章节内容的编排上，我也调整过多次。

我在 2016 年回合肥定居以后，因为工作的原因一直很忙碌，时间基本完全碎片化，很难抽出完整的时间来进行编写和校验，最后一版的校验还是熬了近一周的夜才完成。这里非常感谢本书的另一位作者臧强磊，正是因为他在前期概念梳理和测试上的大量付出，才有了本书，同时也要感谢刘浩在测试上提供的帮助。相信你们在这个过程中也收获很多。

我从 2017 年开始一直在中科大管理学院学习 MBA 课程，这两年课程的学习压力很大，尤其是 2018 年核心课程的学习，这些课程让我学到了很多 IT 技能之外的知识。这里非常感谢中科大 MBA 中心提供了一个非常好的沟通平台；感谢班主任莫鸿芳老师，莫老师不仅在学习上给予了我很多指导，而且在工作上也给我提供了很多帮助；感谢毕功兵、丁斌、徐毅、周垂日、叶五一、唐述毅、张瑞稳、吴剑琳、朱宁、曹苏等所有的授课老师；感谢中科大 MBA 1709 班的所有同学，非常荣幸能在人生最重要的阶段遇见如此可爱的你们。

最后感谢我的夫人对我的支持与理解，正是因为她的付出，我才有时间来完成本书的编写工作。因为时间和个人精力的原因，本书在编写和校验过程中难免会出现一些错误，读者在阅读的过程中如果发现这些错误，请发邮件告知我，邮箱是 ahdba@qq.com，希望本书能给读者们在学习数据库的过程中提供一些帮助。

戴明明/Dave
2019 年 5 月 5 日

序 2

有的时候真的感觉时间过得太快了，还清楚地记得大学刚入学时军训的场景，转眼间工作将近六年了。幸运的是我一直从事数据库方面的工作，有幸结识了很多数据库方面的专家，对我在数据库方面的学习和工作产生了很大的影响。而就在去年，我和几位小伙伴一起翻译的一本由国外几位 Oracle 数据库领域知名作者编写的图书《Oracle 数据库问题解决方案与故障排除手册》出版了。

后来，Dave 邀请我一起写一本关于 Oracle 18c 方面的图书。因为 Oracle 18c 版本中引入了不少新特性，为了更好地理解这些新特性，我开始研读官方文档，然后进行测试。这里我也给广大读者们一个小小的建议，学习一门技术，官方文档是最好的学习资料，务必仔细研读。这次自己写书和翻译书的感觉完全不同，真正体会到了万事开头难的含义。后来，Dave 给了我一些帮助，帮我梳理了章节的框架，我才有了一些思路开始编写本书。经过近大半年的时间，本书的大部分内容才编写完成，后来又进行了多次校验以保证内容的准确性。

从 Oracle 12c 开始，Oracle 逐渐向智能化方向发展，一些日常维护操作无须 DBA 进行干预，尤其是 Oracle 18c 中提出了“自治式”数据库，也就是数据库会自动进行一些日常的运维操作，如升级、打补丁等，并进一步增强了多租户、In-Memory 等技术。由于 Oracle 数据库知识点非常多且内容非常深奥，对于大部分读者来说通过官方文档了解 Oracle 可能有一些难度而且非常耗时，所以我才有了动手编写本书的想法，一方面自己学习，另一方面要把知识传播给广大读者朋友们。

本书重点介绍 Oracle 18c 中引入的重要的新特性，为了详细介绍每个重要的新特性，特意将每个新特性单独作为一章进行讲解。由于篇幅有限，删减了一点内容，留下了特性中最重要的部分，尽可能地让读者理解每个新特性的特点及适用场景。

本书的出版要感谢家人的理解和支持，没有更多的时间陪伴你们；感谢 Dave 对本书内容不厌其烦的校验审核，相比于写书，审核校验更为枯燥乏味；最后感谢一起从事数据库行业的小伙伴们的支持，希望本书可以给你们带来帮助。

由于时间有限，尽管已经利用了很多休息时间去校验，但还可能有不完美的地方，恳请广大读者批评指正。

臧强磊

2019 年 4 月 29 日

读者服务

轻松注册成为博文视点社区用户（www.broadview.com.cn），扫码直达本书页面。

- **下载资源**：本书如提供示例代码及资源文件，均可在 下载资源 处下载。
- **提交勘误**：您对书中内容的修改意见可在 提交勘误 处提交，若被采纳，将获赠博文视点社区积分（在您购买电子书时，积分可用来抵扣相应金额）。
- **交流互动**：在页面下方 读者评论 处留下您的疑问或观点，与我们和其他读者一同学习交流。

页面入口：http://www.broadview.com.cn/35502

目录

第1章 多租户架构

多租户（Multitenant）架构在 Oracle 12cR1 中被引入，经过几年的发展，到现在的 Oracle 18c，多租户已经是一个很成熟的架构。要学习 Oracle 18c，必须先掌握多租户的知识。

在多租户架构下，Oracle 数据库作为多租户容器数据库（CDB: Container DataBase）运行。一个 CDB 可以包括零、1 个或多个可插拔数据库（PDB: Pluggable Database）。每个 PDB 之间是独立的，相互不影响，但它们都接受 CDB 的管理。而在 Oracle 12c 之前的版本中，数据库都是 Non-CDB 架构的。

1.1 CDB 中的容器

容器是多租户架构中数据或元数据的逻辑结合。如图 1-1 所示是 CDB 中的容器架构。

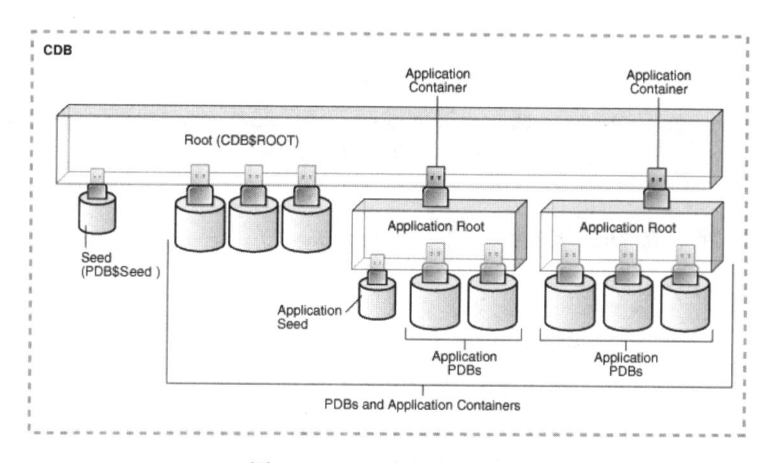

图 1-1 CDB 中的容器架构

一个 CDB 中可以包含以下容器。

- 一个CDB 根容器（简称Root）。它是每个PDB中所属的模式（Schemas）、模式对象和非模式对象的组合。根容器中存储着 Oracle 内部提供的元数据和公共用户，比如 PL/SQL 包就是一种元数据。公共用户对所有容器都可见，只要公共用户有合适的权限，就可以连接和管理所有容器，包括 PDB。
- 一个系统容器（System Container）。系统容器包含根容器和 CDB 中的所有 PDB，它是 CDB 的一个逻辑容器。
- 零或多个应用程序容器（Application Container）。应用程序容器由一个应用程序根容器（Application Root）和多个插到该根容器的 PDB 组成。系统容器包含 CDB 根容器和 CDB 中的所有 PDB，而应用程序容器只包含已插到应用程序根容器的 PDB。应用程序根容器只属于 CDB 根容器，不属于其他容器。
- 零或多个用户创建的 PDB。
- 一个种子 PDB（Seed PDB）。种子 PDB 是系统提供的模板，在 CDB 中可以使用该模板来创建新的 PDB。注意，我们无法对种子 PDB 中的对象进行添加或修改。

1.2 CDB 架构的优点

下面先看一下 Non-CDB 架构中存在的问题。如果公司有数百套数据库，那么 DBA 管理这些数据库的成本是非常高的，而且升级数据库、给数据库打补丁等工作非常耗时，也非常容易出错。近几年来，硬件的性能提升非常明显，特别是 CPU 核数的增加，使服务器的性能大幅度提升。如果一个服务器上只装一个数据库，则会造成硬件和人力资源的严重浪费。

1.2.1 利于数据库整合

在硬件性能足够用的情况下，可以将多个数据库中的数据集中到一个数据库中。Oracle CDB 架构可以在不更改现有模式和应用程序的情况下将数据和代码合并。PDB 和 Non-CDB 对应用程序来说是完全透明的，包括连接方式都是相同的。即使是使用 Oracle DataGuard，在数据库备份和还原时，对 Non-CDB 和 CDB 来说操作都是一样的。

使用 CDB 架构进行数据库整合有以下好处。

- 降低成本：通过将硬件和数据库的相关组件整合到同一组后台进程中，可以有效地共享计算 CPU 和内存资源，从而降低硬件和维护的成本。例如，单台服务器上的 100 个 PDB 共享一个数据库实例。
- 数据和代码的快速迁移：在 CDB 架构中，可以实现将 PDB 快速地插到 CDB 中，先从 CDB 中拔下 PDB，然后将这个 PDB 插到另一个 CDB 中，还可以在线复制 PDB。如果 CDB 的字符集是 AL32UTF8，那么不同数据库字符集的 PDB 可以存储在同一个 CDB 中，即 CDB 中可以插入任何字符集的 PDB，并在不进行字符集转换的情况下使用它。
- 易于管理和监控：DBA 可以通过执行单个操作（如打补丁或执行 RMAN 备份）来管理环境，可以一次性管理 CDB 和 CDB 中所有的 PDB，并且简化了备份策略和灾难恢复。
- 数据和代码的分离：数据库通过整合合并到了一个物理数据库中，而 PDB 和 Non-CDB 的管理非常相似。例如，如果用户误删除了数据，那么 DBA 可以使用 Oracle Flashback 或基于时间点恢复来恢复数据，而不会影响其他 PDB。
- 管理权限的分离：公共用户只要有足够的权限就可以连接到任意一个 PDB，而本地用户只能连接到指定的 PDB。管理员将职责划分如下。
    - 管理员使用公共用户管理 CDB 或应用程序容器。由于权限只包含在授予它的 PDB 中，因此一个 PDB 上的本地用户对同一个 CDB 中的其他 PDB 没有访问权限。
    - 管理员使用本地用户管理单个 PDB。

注意：公共用户和本地用户是 Oracle 12c 中引入的概念，公共用户只能在 CDB 中创建，用来管理 CDB 中所有的 PDB，默认情况下公共用户的名称只能以 C##开头。本地用户只能访问指定的 PDB，而无法访问其他的 PDB。

- 简化性能调优：很明显，收集一个数据库的数据要比收集多个数据库的数据更容易。例如，修改一个数据库 SGA 的大小比修改 100 个数据库 SGA 的大小简单。
- 简化数据库补丁和升级：采用 CDB 架构，可以减少需要升级的数据库的数量，减少出错的机会。

1.2.2 利于数据库管理

在 CDB 架构中，PDB 中也会存储本身的数据和元数据，而不是将所有 PDB 的字典元数据都存储在同一个数据库中。通过分开存储各自的字典元数据，PDB 可以更容易地作为一个独立的单元进行管理。这种分开存储有以下好处。

- 更容易升级数据和代码：例如，可以不再升级 CDB，而是直接将 PDB 从现有的 CDB 中拔出，然后插到更高版本的 CDB 中。
- 更容易在服务器之间迁移：执行负载平衡或满足 SLA 时，可以将应用程序数据库从本地数据中心迁移到云，或者在同一个环境中的两个服务器之间迁移。
- 防止 PDB 中的数据损坏：可以将某个 PDB 闪回到特定的 SCN 或 PDB 的恢复点，而不影响其他 PDB。这个特性类似于 Non-CDB 的闪回数据库特性。

1.3 多租户存储的物理结构

1.3.1 数据文件的存放位置

因为后面章节的内容涉及多租户存储的物理结构，所以这里提前说明。在多租户架构中，与数据库相关的各个文件的存放位置和 Oracle 11g 中相关文件的存放位置有明显区别，主要体现在 PDB 的数据文件上，而其他的文件，如控制文件、Redo 日志文件、密码文件、参数文件都没有变化。

如果 PDB 的数据文件使用 Oracle 自动管理（OMF），那么文件存放在 $ORACLE_BASE/oradata/SID/DATAFILE/PDB_GUID/目录中。通过手动方式管理 PDB 的数据文件，理论上数据文件的存放位置是没有限制的，但是为了方便管理，建议参考标准去存放，将 PDB 相关的数据文件存放在单独的目录中。

CDB 和 Non-CDB 有着相似的结构，除了每个 PDB 都有自己的表空间，如 SYSTEM 表空间、SYSAUX 表空间、TEMP 表空间和 UNDO 表空间（只有本地 UNDO 模式才有，如果是共享 UNDO 模式，则所有 CDB Root 和 PDB 共享一个 UNDO 表空间）。

注意：默认情况下，通过 DBCA 创建的 CDB 都会启动本地 UNDO 模式。

下面分别说明本地 UNDO 模式和共享 UNDO 模式下的 CDB 结构。

- 本地 UNDO 模式：实际上除在线 Redo 日志文件外，其他表空间的每个 PDB 中都有。所有 PDB 会共享 CDB 的在线 Redo 日志文件，如图 1-2 所示。

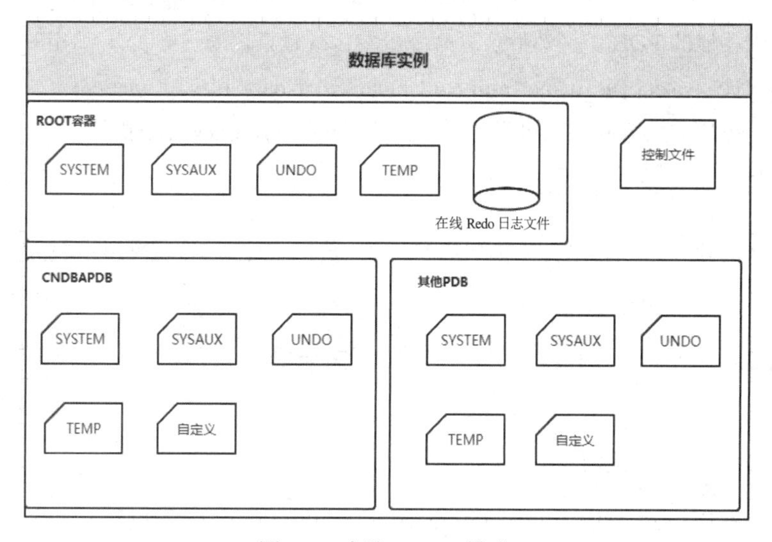

图 1-2　本地 UNDO 模式

- 共享 UNDO 模式：和本地 UNDO 模式唯一的区别就是，所有 PDB 共享一个 UNDO 表空间，如图 1-3 所示。

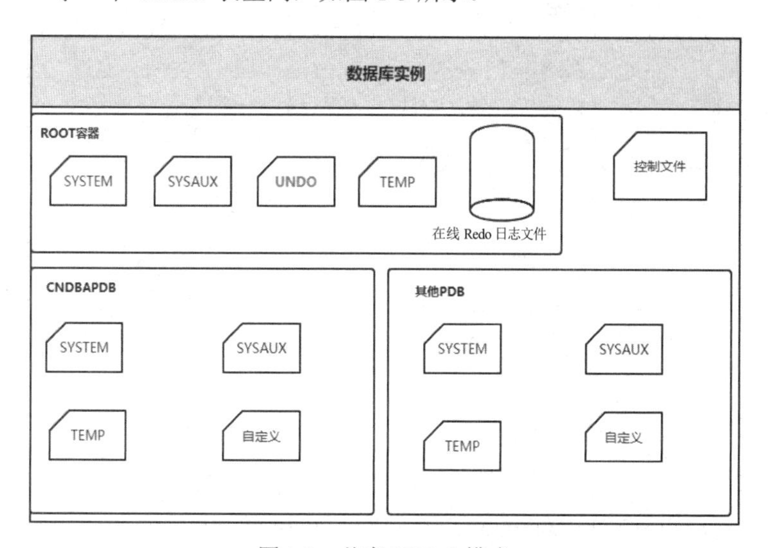

图 1-3　共享 UNDO 模式

1.3.2 查看 CDB 数据文件目录

查看 CDB 数据文件的存放位置，需要切换到相应的容器下。例如，查看某个 PDB 数据文件的存放目录，可以通过 alter session set container=pdbname 命令切换到相应的 PDB 下，然后执行命令查看；或者直接通过 CONTAINER 子句来查询，如“select file_name from containers（dba_data_files） where con_id=3;”，这里的容器 ID（con_id）可通过 v$pdbs 视图或 show pdbs 命令进行查看。

- 单实例文件系统。

  由于没有启动 OMF，所以所有的数据文件命名及存放目录都由 DBA 手动指定。

```
SQL> show parameter db_create_file_dest
NAME                          TYPE       VALUE
------------------------------------------------------------
db_create_file_dest               string
```

- CDB root

```
SQL> show con_name
CON_NAME
------------------------------
CDB$ROOT
SQL> select file_name from dba_data_files;
FILE_NAME
------------------------------------------------------
/u01/app/oracle/oradata/LEI/system01.dbf
/u01/app/oracle/oradata/LEI/sysaux01.dbf
/u01/app/oracle/oradata/LEI/undotbs01.dbf
/u01/app/oracle/oradata/LEI/users01.dbf
```

- PDB

  切换到相应的 PDB 下进行查询即可。

```
SQL> show pdbs;
   CON_ID CON_NAME                      OPEN MODE
RESTRICTED
---------- ------------------------------ ----------
2 PDB$SEED                    READ ONLY  NO
       3 LEIPDB                        READ WRITE NO
SQL> alter session set container=leipdb;
Session altered.
```

可以看到，CDB root 数据文件所在的目录中创建了一个子目录，该子目录用于存放相应 PDB 的数据文件。

```
SQL> select file_name from dba_data_files;
FILE_NAME
-------------------------------------------------------
/u01/app/oracle/oradata/LEI/leipdb/system01.dbf
/u01/app/oracle/oradata/LEI/leipdb/sysaux01.dbf
/u01/app/oracle/oradata/LEI/leipdb/undotbs01.dbf
/u01/app/oracle/oradata/LEI/leipdb/users01.dbf
```

- RAC 环境下的 ASM 存储。

为了进行对比，在 RAC 环境中启用了 OMF。

```
SQL> show parameter db_create_file_dest
NAME                  TYPE       VALUE
------------------------------------ -----------
db_create_file_dest        string     +DATA
```

- CDB root

```
SQL> select file_name from dba_data_files;
FILE_NAME
-------------------------------------------------------
+DATA/CNDBA/DATAFILE/system.258.984927033
+DATA/CNDBA/DATAFILE/sysaux.257.984926993
+DATA/CNDBA/DATAFILE/undotbs1.260.984927081
+DATA/CNDBA/DATAFILE/users.259.984927079
+DATA/CNDBA/DATAFILE/undotbs2.268.984927511
```

- PDB

可以看到，Oracle 使用 PDB 的 GUID 来区分不同 PDB 数据文件存放的目录。

```
SQL> show con_name
CON_NAME
------------------------------
CNDBAPDB
SQL> select file_name from dba_data_files;
FILE_NAME
-------------------------------------------------------
+DATA/CNDBA/7415D8FAA2156610E0533701A8C09D17/DATAFILE/
system.273.984927777
+DATA/CNDBA/7415D8FAA2156610E0533701A8C09D17/DATAFILE/
sysaux.272.984927777
```

```
+DATA/CNDBA/7415D8FAA2156610E0533701A8C09D17/DATAFILE/
users.275.984927829
+DATA/CNDBA/7415D8FAA2156610E0533701A8C09D17/DATAFILE/
test.276.984929503
```

1.4 创建 CDB

Oracle 18c 的软件安装和之前版本的软件安装有非常大的区别。之前版本解压缩的安装文件可以存放在任何位置，而 Oracle 18c 的安装文件会解压缩到 Oracle home，可以从 Oracle home 目录中运行 runInstaller 命令。

```
    [root@cndba software]# chown oracle:oinstall LINUX.X64_
180000_db_home.zip
    [root@cndba software]# su - oracle
    [oracle@cndba software]$ ll
    total 4457668
    -rw-r--r--. 1 oracle oinstall 4564649047 Jul 30 10:14
LINUX.X64_180000_db_home.zip
    [oracle@cndba software]$ unzip -d /u01/app/oracle/product/
18.1.0/dbhome_1/ LINUX.X64_180000_db_home.zip
```

切换到 Oracle 用户，进入 Oracle home 目录，执行./runInstaller 命令安装 Oracle 18c 数据库软件。

```
    [oracle@cndba dbhome_1]$ cd /u01/app/oracle/product/18.1.0/
dbhome_1/
    [oracle@cndba dbhome_1]$ ./runInstaller
```

因为篇幅有限，完整的 Oracle 18c 的安装步骤可以参考笔者的博客文章：《Oracle 18c 单实例安装手册 详细截图版》（https://www.cndba.cn/dave/article/2971）。

在 Oracle18c 软件安装完成之后，可以通过 DBCA 命令创建 CDB，也可以手动创建 CDB。在 DBCA 的 GUI 中选择“Create a database”单选按钮来创建一个新的数据库实例，如图 1-4 所示。

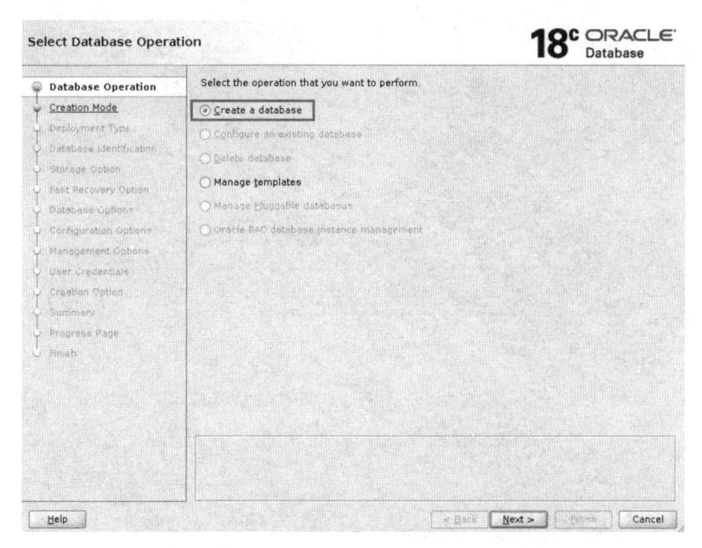

图 1-4　选择“Create a database”单选按钮

单击“Next”按钮，在打开的对话框中选择“Advanced configuration”单选按钮，用高级选项配置实例的相关参数，如图 1-5 所示。

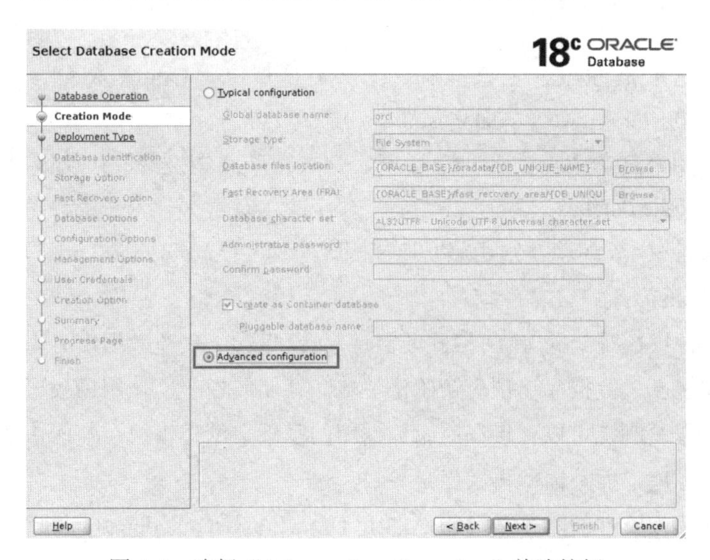

图 1-5　选择“Advanced configuration”单选按钮

单击“Next”按钮，在打开的对话框中根据实际情况选择数据库类型和模板名称，如图 1-6 所示。

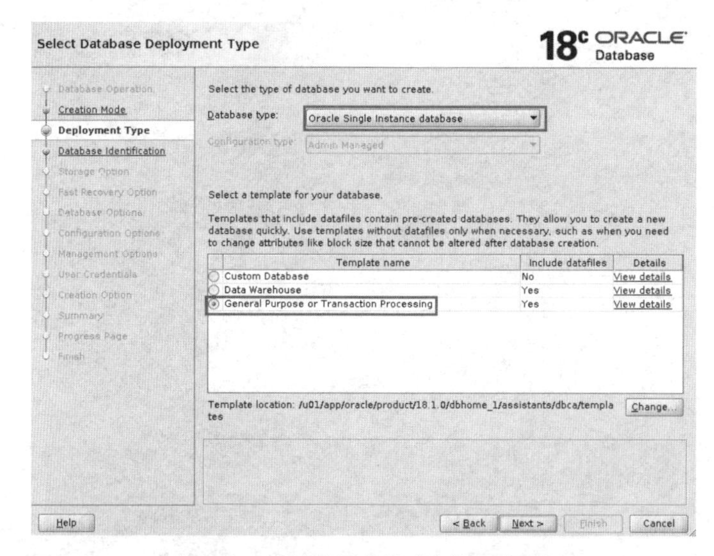

图 1-6　选择数据库类型和模板名称

数据库类型的选择方法如下。

- 如果是单实例环境，则数据库类型选择“Oracle Single Instance database”。
- 如果是 RAC 环境，则数据库类型选择“Oracle Real Application Cluster（RAC） database”。
- 如果是单节点的 RAC 环境，则数据库类型选择“Oracle RAC one Node database”。

如果是 RAC 环境，则还需要选择配置类型（Configuration type），默认为“Admin Managed”。

DBCA 模板是一些 XML 格式的文件集合，包含了创建数据库必需的信息。这些信息包括：数据库选项、初始化参数、存储属性等。使用模板可以节省时间，也利于复制，可以直接在原始数据库的基础上创建模板，从而快速创建一个复制数据库，而不需要再次指定参数。

默认模板文件的路径是 $ORACLE_HOME/assistants/dbca/templates。

```
[oracle@18c ~]$ ll $ORACLE_HOME/assistants/dbca/templates
total 422220
-rw-r-----. 1 oracle oinstall 4888 Feb 8 09:20 Data_Warehouse.dbc
-rw-r-----. 1 oracle oinstall 4768 Feb 8 09:20 General_Purpose.dbc
-rw-r-----. 1 oracle oinstall 10772 Sep 10  2017 New_Database.dbt
```

```
    -rw-r-----. 1 oracle oinstall 124870656 Feb 8 08:32 pdbseed.dfb
    -rw-r-----. 1 oracle oinstall 6643 Feb 8 08:32 pdbseed.xml
    -rw-r-----. 1 oracle oinstall 18726912 Feb 8 09:19
Seed_Database.ctl
    -rw-r-----. 1 oracle oinstall 288718848 Feb 8 09:19
Seed_Database.dfb
```

Oracle 提供的 DBCA 模板包括 Data Warehouse、General Purpose or Transaction Processing 和 Custom Database。通常选择“General Purpose or Transaction Processing”模板。

根据安装要求，输入 Global database name 和 SID，然后选择是否创建容器数据库（“Create as Container database”复选框）。由于创建的是 CDB，所以这里必须选择创建容器数据库（如果是 Non-CDB，则不需要勾选 “Greate as Container database” 复选框）。选择创建 PDB 的数量并设置 PDB 的名称，如图 1-7 所示。PDB name 表示 PDB 名称的前缀，如果创建多个 PDB，则后面还会加上数字。例如，创建两个 PDB，将 PDB name 设置为 CNDBA，那么最终的 PDB 名称分别是 CNDBA1 和 CNDBA2。

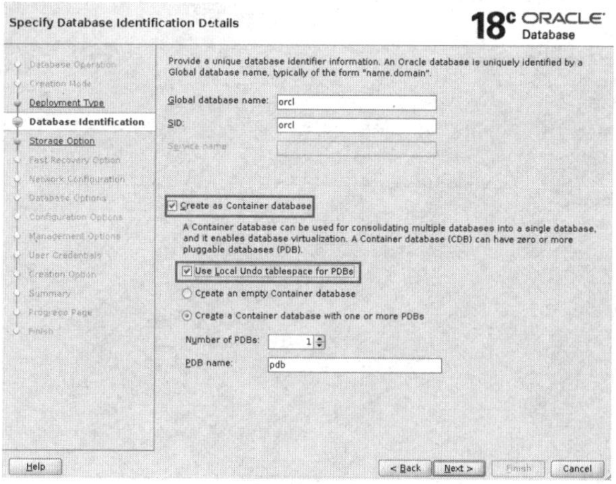

图 1-7　设置创建选项

单击“Next”按钮，在打开的对话框中设置数据库存储选项，默认使用模板的存储选项，也可以自己设置存储选项，包括 ASM、文件系统、数据文件存放路径、是否使用 OMF 等，如图 1-8 所示。

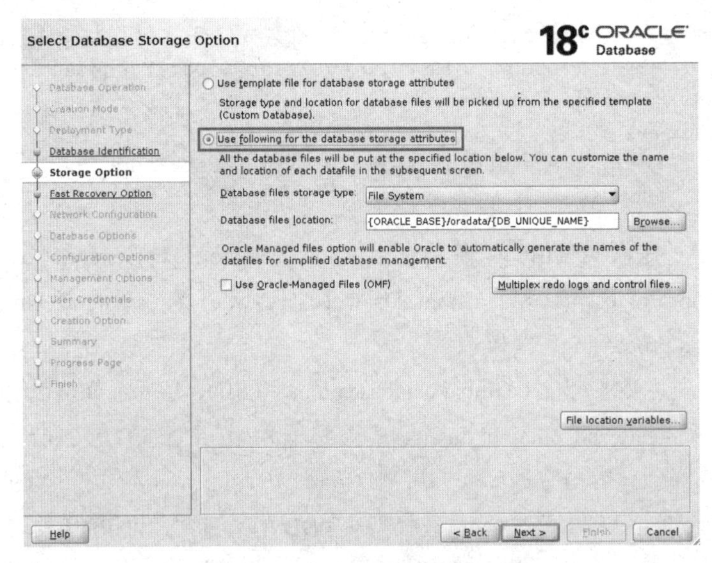

图 1-8　设置数据库存储选项

单击“Next”按钮，在打开的对话框中设置快速恢复选项，包括路径、大小、ASM、文件系统、是否启用归档等，如图 1-9 所示。

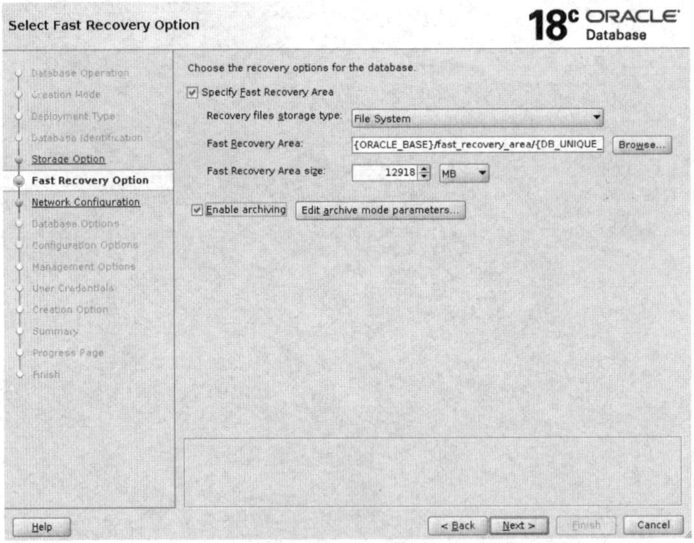

图 1-9　设置快速恢复选项

单击“Next”按钮，在打开的对话框中设置网络选项，可以选择已有的监听，也可以创建新的监听，如图 1-10 所示。

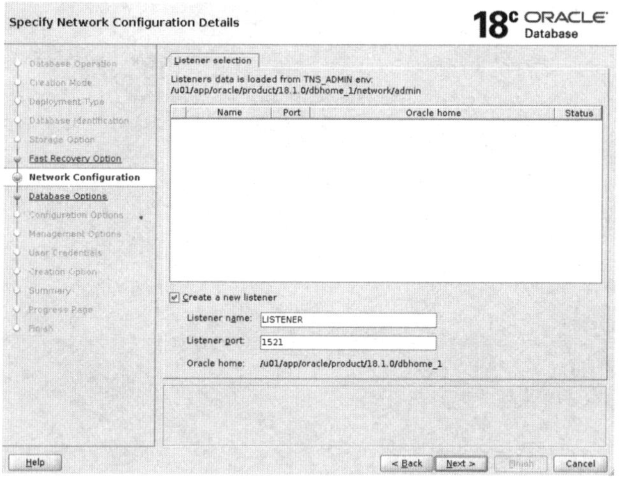

图 1-10　设置网络选项

单击“Next”按钮，在打开的对话框中设置数据仓库选项。数据仓库（Data Vault）是 Oracle 数据库的安全组件之一，它的主要功能是保护敏感数据，可以根据需要进行设置，如图 1-11 所示。关于数据仓库的更多内容可以参考官方手册。

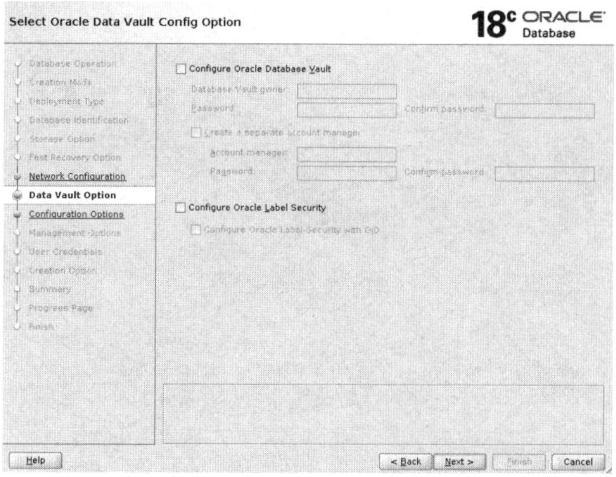

图 1-11　设置数据仓库选项

单击“Next”按钮，在打开的对话框中设置配置选项，这里的配置选项比较多，包括内存（SGA、PGA、管理方式）、进程数、字符集、连接模式、图式示例，如图 1-12 所示。

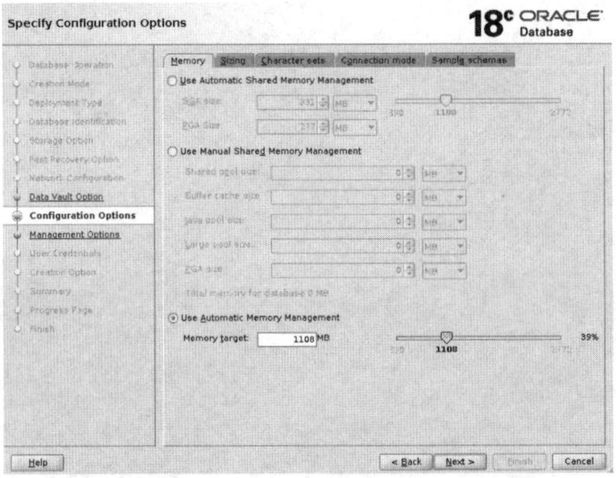

图 1-12　设置配置选项

单击“Next”按钮，在打开的对话框中设置 OEM 选项，如图 1-13 所示。在 Oracle 10g 和 Oracle 11g 版本中，在建库时一般都不安装 OEM，因为收集信息会消耗部分资源，而且大部分 DBA 没有使用 OEM 来查看数据库状态的习惯。随着 Oracle 13c OEM 的发布，OEM 的功能越来越强大，如果 DBA 管理的数据库很多，就可以选择独立的服务器来安装 Oracle 13c 的 OEM Cloud Control，数据库服务器端的 OEM 还是不建议安装。

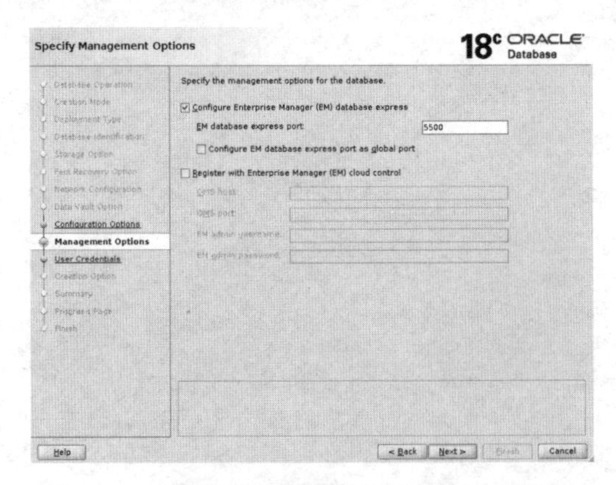

图 1-13　设置 OEM 选项

单击“Next”按钮，在打开的对话框中设置数据库管理员的密码，根据安全规范设置密码，尽量避免使用简单的密码，如图 1-14 所示。

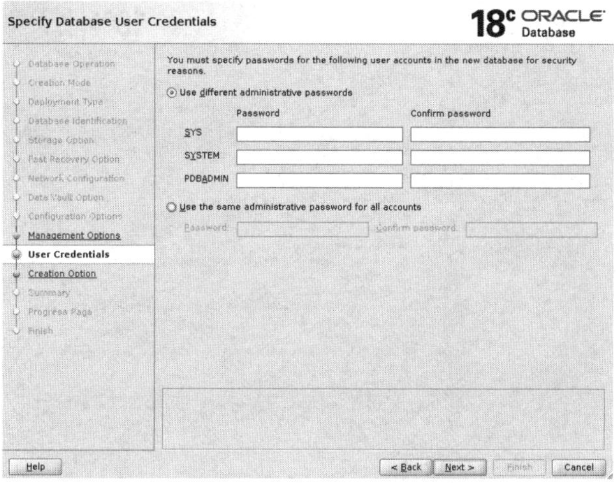

Specify Database User Credentials　18c ORACLE Database

Database Operation
Creation Mode
Deployment Type
Database Identification
Storage Option
Fast Recovery Option
Network Configuration
Data Vault Option
Configuration Options
Management Options
User Credentials
Creation Option
Summary
Progress Page
Finish

You must specify passwords for the following user accounts in the new database for security reasons.

Use different administrative passwords

	Password	Confirm password
SYS		
SYSTEM		
PDBADMIN		

Use the same administrative password for all accounts

Password:　Confirm password:

Help　< Back　Next >　Finish　Cancel

图 1-14　设置数据库管理员的密码

单击“Next”按钮，在打开的对话框中设置数据库创建选项，包括是否创建数据库、是否生成创建数据库的脚本（可以手动调用脚本创建）、是否保存这个数据库的配置模板（以后可以直接使用该模板创建数据库），如图 1-15 所示。

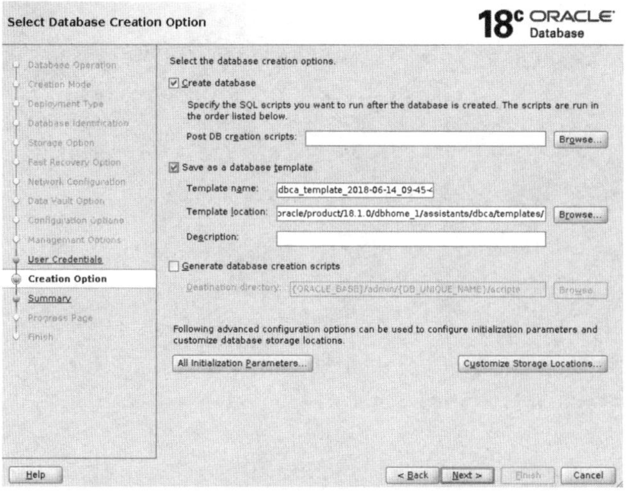

Select Database Creation Option　18c ORACLE Database

Database Operation
Creation Mode
Deployment Type
Database Identification
Storage Option
Fast Recovery Option
Network Configuration
Data Vault Option
Configuration Options
Management Options
User Credentials
Creation Option
Summary
Progress Page
Finish

Select the database creation options.

Create database

Specify the SQL scripts you want to run after the database is created. The scripts are run in the order listed below.

Post DB creation scripts:　Browse...

Save as a database template

Template name: dbca_template_2018-06-14_09-45-

Template location: oracle/product/18.1.0/dbhome_1/assistants/dbca/templates/　Browse...

Description:

Generate database creation scripts

Destination directory: {ORACLE_BASE}/admin/{DB_UNIQUE_NAME}/scripts　Browse

Following advanced configuration options can be used to configure initialization parameters and customize database storage locations.

All Initialization Parameters...　Customize Storage Locations...

Help　< Back　Next >　Finish　Cancel

图 1-15　设置数据库创建选项

单击“Next”按钮，在打开的对话框中检查数据库配置是否有误，确认无误后，单击“Finish”按钮开始创建数据库，如图 1-16 所示。

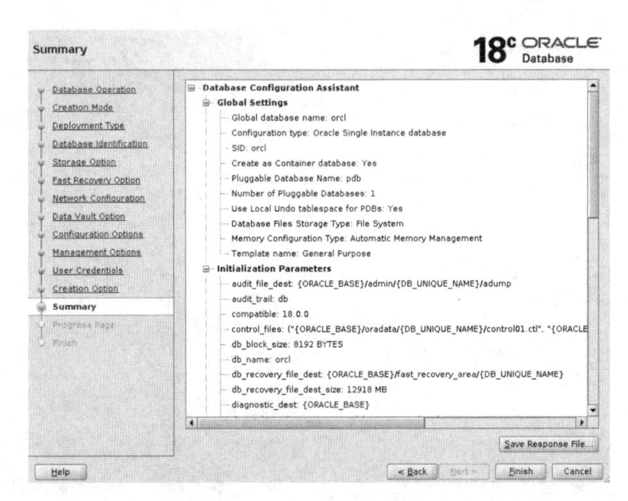

图 1-16 检查数据库配置选项

第2章

PDB

PDB 是运行在 CDB 上的一个数据库，各个 PDB 之间独立运行。在 CDB 中创建、删除、迁移 PDB 非常方便，不会对 CDB 中的其他 PDB 有任何影响。

2.1 创建 PDB 的常用方法

创建 PDB 的常用方法如图 2-1 所示。

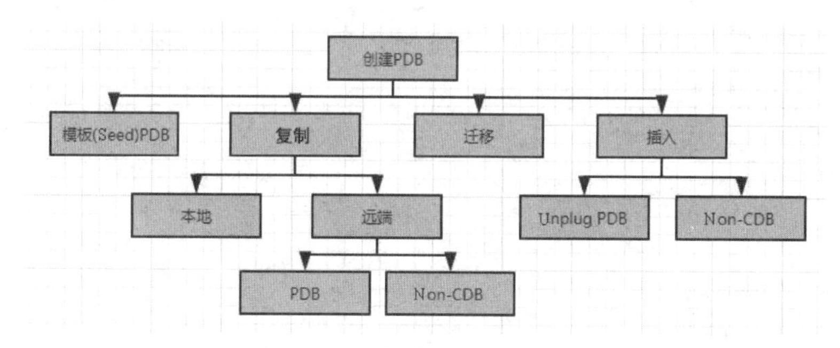

图 2-1 创建 PDB 的常用方法

下面对创建 PDB 的常用方法做简单的说明，如表 2-1 所示。

表 2-1 创建 PDB 的常用方法的简单说明

创建 PDB 的方法	说 明
利用 Seed（种子）模板来创建	默认创建 PDB 的方法。使用 PDB Seed 或应用程序 Seed 的模板文件来创建 PDB。将模板 PDB 相关的文件复制到一个新的目录，并将新的文件与新的 PDB 进行关联
复制已有的 PDB/Non-CDB	通过复制 PDB 或 Non-CDB 来创建 PDB。源库可以是本地 CDB 中的 PDB、远端 CDB 中的 PDB、本地或远端应用程序容器中的 PDB、Non-CDB。该方法将与源端相关的文件复制到新的目录，然后与新的 PDB 进行关联

续表

创建 PDB 的方法	说　明
将 PDB 迁移到另一个 CDB	通过将 PDB 从一个 CDB 中移动到另一个 CDB 中创建 PDB。该方法将与 PDB 相关的文件移动到一个新的目录，而不是复制
将 Unplug 的 PDB 插入到 CDB 中	通过使用 PDB 的 XML 元数据文件来创建 PDB，并将其插到 CDB 中。XML 文件中记录了 PDB 的详细配置信息
从 Non-CDB 创建 PDB，并插入到 CDB 中	利用 DBMS_PDB 包将 Non-CDB 创建为 PDB 并插到 CDB 中

2.2 创建 PDB 的参数说明

如表 2-2 所示是创建 PDB 时可以使用的参数的说明。

表 3-2 创建 PDB 时可以使用的参数的说明

参　数	说　明	使 用 情 况
AS APPLICATION CONTAINER	是否创建 APPLICATION 容器	在 CDB 中创建应用程序容器
AS CLONE	是否将一个 PDB 插到 CDB 中	将一个 Unplug PDB 插到一个 CDB 中
AS SEED	是否在应用程序容器中创建应用程序 Seed（种子）	在应用程序容器中创建应用程序 Seed
CREATE_FILE_DEST	是否启用 OMF 来管理 PDB，如果不启用 OMF，则需要设置下面三个选项中的一个： • 使用 FILE_NAME_CONVERT； • CDB 启动 OMF； • 使用 PDB_FILE_NAME_CONVERT	• 利用 Seed 模板创建 PDB。 • 复制 PDB。 • 迁移 PDB。 • 将一个 Unplug PDB 插到一个 CDB 中
DEFAULT TABLESPACE	是否为 PDB 指定默认的永久表空间	• 利用 Seed 模板创建 PDB。 • 复制 PDB。 • 迁移 PDB。 • 将一个 Unplug PDB 插到一个 CDB 中
FILE_NAME_ CONVERT	指定目标端文件的存放位置。如果不使用 FILE_NAME_CONVERT 参数，则需要设置下面三个选项中的一个： • 使用 CREATE_FILE_DEST；	• 利用 Seed 模板创建 PDB。 • 复制 PDB。 • 迁移 PDB。 • 将一个 Unplug PDB 插到一个 CDB 中

续表

参　数	说　明	使用情况
	• CDB 启动 OMF； • 使用 PDB_FILE_NAME_CONVERT	
COPY/MOVE	是否复制、移动文件到新的目录（不启用 NOCOPY），与下面四个选项中的一个选项结合使用： • 指定 FILE_NAME_CONVERT； • 指定 CREATE_FILE_DEST； • 启用 OMF； • 指定 PDB_FILE_NAME_CONVERT	将一个 Unplug PDB 插到一个 CDB 中
NO DATA	只复制源 PDB 的对象定义，不复制数据	复制 PDB
PARALLEL	指定创建 PDB 的并行数	• 利用 Seed 模板创建 PDB。 • 复制 PDB
REFRESH MODE MANUAL 或 REFRESH MODE EVERY minutes	是否能够刷新 PDB 以将更改的数据从源 PDB 传播到复制 PDB。必须以只读模式打开可刷新的 PDB（禁用可刷新 PDB：不指定子句或设置 REFRESH MODE NONE）	复制 PDB
SERVICE_NAME_CONVERT	重命名新 PDB 的 SERVICE NAME	• 利用 Seed 模板创建 PDB。 • 复制 PDB。 • 迁移 PDB。 • 将一个 Unplug PDB 插到一个 CDB 中
SNAPSHOT COPY	是否使用存储快照来复制 PDB（要求：底层的文件系统支持存储快照）	复制 PDB
SNAPSHOT MODE MANUAL 或 SNAPSHOT MODE EVERY [MINUTES|HOURS]	是否启用 PDB 快照	• 利用 Seed 模板创建 PDB。 • 复制 PDB。 • 迁移 PDB。 • 将一个 Unplug PDB 插到一个 CDB 中
STORAGE	指定 PDB 可以使用的空间大小	• 利用 Seed 模板创建 PDB。 • 复制 PDB。 • 迁移 PDB。 • 将一个 Unplug PDB 插到一个 CDB 中

续表

参　　数	说　　明	使用情况
TEMPFILE REUSE	是否重新使用目标端已存在的临时文件	• 利用 Seed 模板创建 PDB。 • 复制 PDB。 • 迁移 PDB。 • 将一个 Unplug PDB 插入到一个 CDB 中
USER_TABLESPACES	指定新的 PDB 将包含哪些表空间	• 复制 PDB。 • 将一个 Unplug PDB 插入到一个 CDB 中
ROLES	将指定的角色赋给 PDB_DBA 角色	利用 Seed 模板创建 PDB
USING SNAPSHOT	基于 PDB 快照来创建 PDB	复制一个 PDB 快照

2.3 创建 PDB 必须满足的条件

创建 PDB 必须满足的条件如下：

- CDB 必须存在且处于读写模式；
- 当前用户必须是公共用户，并且当前的容器必须是 CDB root 或应用程序容器；
- 当前用户必须有 CREATE PLUGGABLE DATABASE 权限；
- PDB 名称不能和已存在的 PDB 名称重复；
- 其他一些限制，例如，如果在 DataGuard 环境下创建 PDB，则必须进行更多的额外配置等。

2.4 利用 Seed 模板创建 PDB

下面的示例利用 Seed 模板创建 PDB，代码如下。

```
SQL > CREATE PLUGGABLE DATABASE cndbapdb
 ADMIN USER cndbaadm IDENTIFIED BY testpwd
  ROLES= (DBA)
 STORAGE (MAXSIZE 2G)
 DEFAULT TABLESPACE cndba
  DATAFILE '/u01/app/oracle/oradata/cndbapdb/cndba01.dbf' SIZE
100M AUTOEXTEND ON
  PATH_PREFIX = '/u01/app/oracle/oradata/cndbapdb/'
 FILE_NAME_CONVERT =
('/u01/app/oracle/oradata/pdbseed','/u01/app/oracle/oradata/cndbap
db');
```

```
Pluggable database created

SHOW PDBS;

   CON_ID CON_NAME           OPEN MODE  RESTRICTED
------------------------------------------------------------
    2 PDB$SEED           READ ONLY  NO
    3 CNDBAPDB           MOUNTED
```

上面的示例创建了一个名为 cndbapdb 的 PDB 和一个具有 PDB_DBA 角色的 cndbaadm 管理用户，将 DBA 权限赋给 PDB_DBA 角色，限制 PDB 可以使用的空间大小为 2GB，创建的表空间为 cndba，并设置为默认表空间。

2.5　复制 PDB

复制 PDB 就是用已经存在的数据库为模板创建一个结构和数据（可以不复制数据）相同的 PDB。在复制期间，如果想要保持源数据库正常运行，则源 CDB 必须处于归档模式且是本地 UNDO 模式。以下数据库可以作为复制对象：

- 本地 PDB；
- 远端 PDB；
- Non-CDB。

2.5.1　复制本地 PDB

可以使用命令和 DBCA 来复制本地 PDB。

1．使用 CREATE PLUGGABLE DATABASE 命令创建 PDB

使用 CREATE PLUGGABLE DATABASE 命令创建 PDB 的语句如下：

```
    CREATE PLUGGABLE DATABASE cndbapdb2 FROM cndbapdb
FILE_NAME_CONVERT = ('/u01/app/oracle/oradata/cndbapdb',
'/u01/app/oracle/oradata/cndbapdb2');
```

这里需要注意以下两点。

- 如果 CDB 没有启动本地 UNDO 模式，则源 PDB 必须以只读模式打开；如果 CDB 是本地 UNDO 模式，则可以以读写模式打开。
- 如果没有启用 OMF，则需要指定 FILE_NAME_CONVERT 参数。

如果不想复制源 PDB 中表的数据，又想限制 PDB 可使用的总空间大小，则可以使用如下语句：

```
    CREATE PLUGGABLE DATABASE cndbapdb2 FROM cndbapdb
FILE_NAME_CONVERT =
('/u01/app/oracle/oradata/cndbapdb','/u01/app/oracle/oradata/cndba
pdb2')
      STORAGE (MAXSIZE 3G)
      NO DATA;
```

2. 利用本地 PDB 快照创建 PDB

（1）手动创建快照

确认 SNAPSHOT 模式为 MANUAL，语句如下：

```
    SELECT SNAPSHOT_MODE "S_MODE", SNAPSHOT_INTERVAL/60
"SNAP_INT_HRS" FROM DBA_PDBS;
    S_MODE       SNAP_INT_HRS
    ------------ ------------
    MANUAL
```

连接到 PDB，语句如下：

```
    alter session set container=cndbapdb;
    Session altered.
```

创建 SNAPSHOT，语句如下：

```
    ALTER PLUGGABLE DATABASE SNAPSHOT cndbapdb_snap0614;
    Pluggable database altered.
```

（2）利用前面创建的快照创建 PDB

```
    CREATE PLUGGABLE DATABASE cndbapdb_copy0614 FROM cndbapdb USING
SNAPSHOT cndbapdb_snap0614;
    Pluggable database created.

    show pdbs;
       CON_ID CON_NAME          OPEN MODE  RESTRICTED
    ------------------------------------------------------------
         2 PDB$SEED          READ ONLY  NO
         3 CNDBAPDB          READ WRITE NO
         4 CNDBAPDB_COPY0614          MOUNTED
```

注意：这里只能使用 OMF 方式来创建 PDB，不能使用参数 FILE_NAME _CONVERT。

3. 利用 DBCA 复制 PDB

在 Oracle 18c 中，可以通过 DBCA 来复制本地的 PDB。下面以 CNDBAPDB 为模板创建一个新的 CNDBAPDB2，如图 2-2~图 2-4 所示。

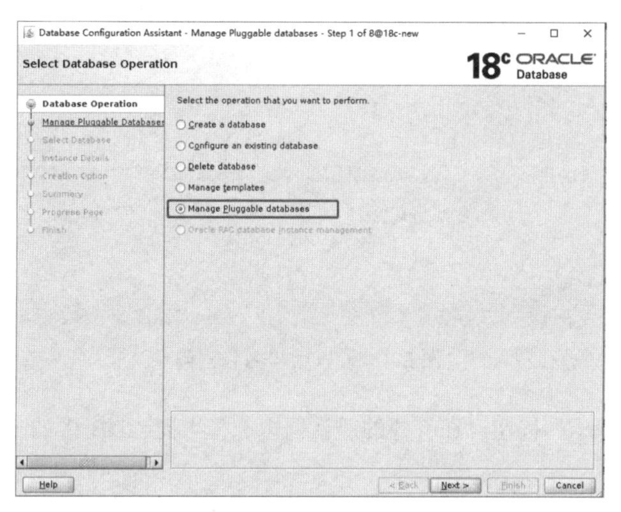

图 2-2　选择“Manage Pluggable database”单选按钮

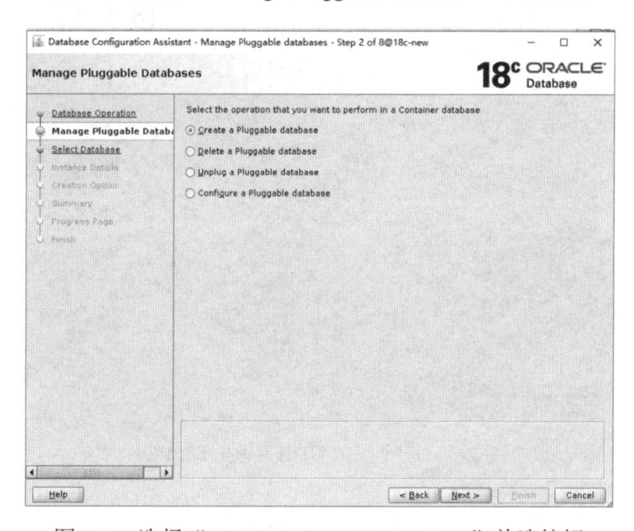

图 2-3　选择“Create a Pluggable database”单选按钮

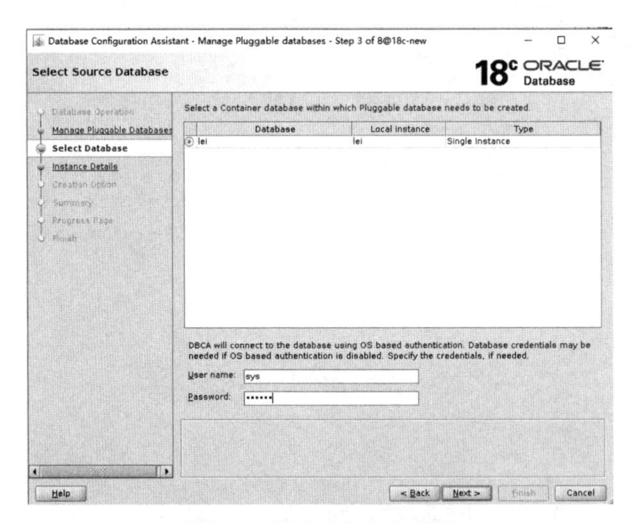

图 2-4　设置用户名和密码

单击“Next”按钮，在打开的对话框中选择当前 CDB 中的一个 PDB，如图 2-5 所示。

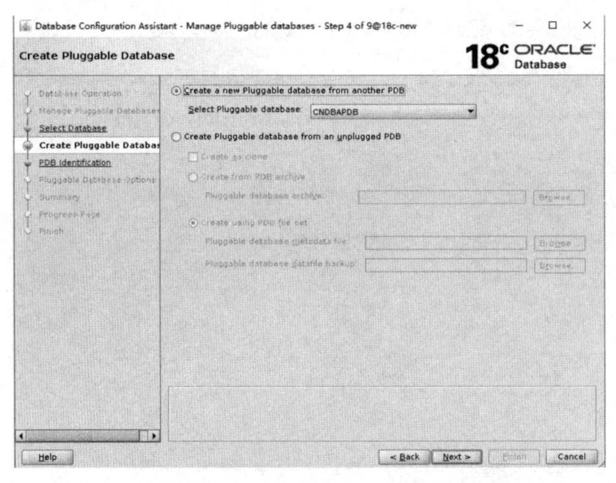

图 2-5　选择当前 CDB 中的一个 PDB

单击“Next”按钮，在打开的对话框中输入新的 PDB 名称，如图 2-6 所示。

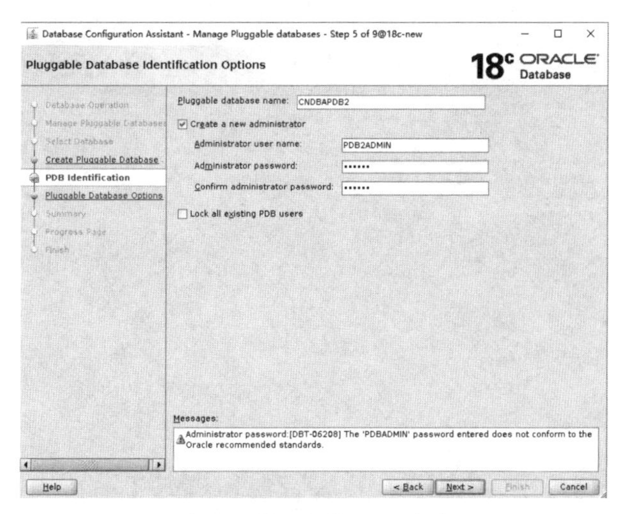

图 2-6　输入新的 PDB 名称

单击“Next”按钮，在打开的对话框中选择数据文件的存储方式及目录，如图 2-7 所示。

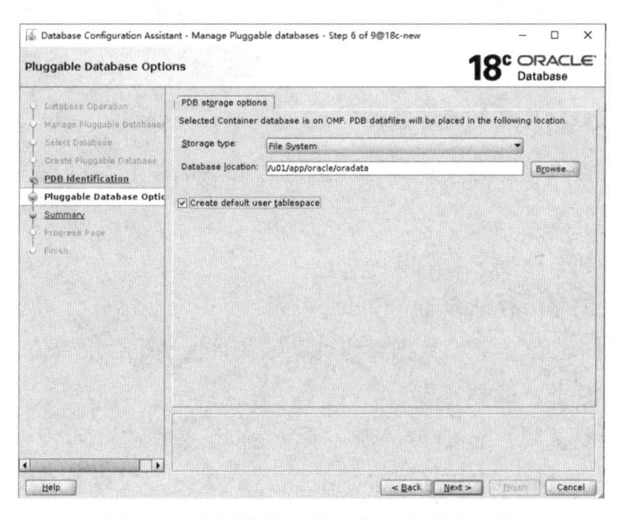

图 2-7　选择数据文件的存储方式及目录

单击“Next”按钮，在打开的对话框中单击“Finish”按钮开始复制，如图 2-8 所示。

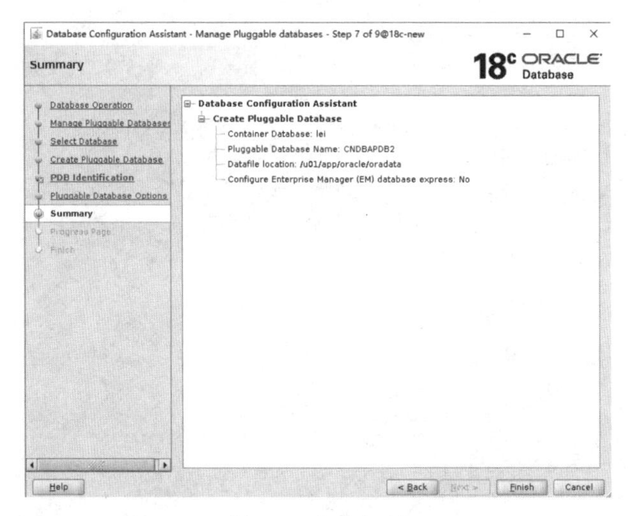

图 2-8　完成复制

2.5.2　复制远端 PDB

如果复制的是远端的 PDB，则必须使用 DBlink 来复制，并且 DBlink 是存在于本地 CDB 中的，而不是存在于远端 CDB 中的。

1. 复制远端 PDB 的必要条件

（1）当前操作用户必须具有 CREATE PLUGGABLE DATABASE 的系统权限。

（2）源端和目标端平台必须满足：

- 同样的字节存储顺序（Endian）；
- 源端平台上安装的数据库组件必须和目标端平台上安装的数据库组件是一样的，或者是其子集。

（3）如果是复制 Non-CDB，则必须满足以下条件：

- CDB 和 Non-CDB 必须运行在 Oracle 12.1.0.2 及以上版本上；
- Blocksize（块大小）必须一致。

（4）如果创建一个应用程序 PDB，则源 PDB 的应用程序名称和版本必须与目标应用程序容器的应用程序名称和版本相匹配。

2. 字符集的要求

- 如果复制 PDB 的 CDB 的字符集不是 AL32UTF8，则源和目标必须具有兼容的字符集和国家字符集；如果复制 PDB 的 CDB 的字符集是 AL32UTF8，则没有这个限制。
- 如果创建一个应用程序 PDB，则应用程序 PDB 必须具有与应用程序容器相同的字符集和国家字符集。同样，如果 CDB 的数据库字符集是 AL32UTF8，则应用程序容器的字符集和国家字符集可以与 CDB 不同。否则，应用程序容器中的所有应用程序 PDB 必须具有与应用程序容器相匹配的相同的字符集和国家字符集。

3. 源 PDB 或 Non-CDB 的模式要求

- 源 PDB 必须处于 Open 状态。
- 在复制 PDB 时，如果远端 CDB 不是本地 UNDO 模式，则源 PDB 必须处于只读模式；如果远端 CDB 是本地 UNDO 模式，则没有此限制。
- 如果远端 CDB 或 Non-CDB 是非归档模式，则源 PDB 或 Non-CDB 必须是只读模式。
- 如果创建的是可刷新的 PDB，则源 PDB 必须是归档模式，并且是本地 UNDO 模式。

4. 操作示例

（1）在主库、备库的 CDB 中执行如下命令，创建相同的共有用户及密码：

```
SQL> create user c##dave identified by oracle;
SQL> grant create session, resource, create any table, unlimited
tablespace to c##dave container=all;
SQL> grant create pluggable database to c##dave container=all;
SQL> grant sysoper to c##dave container=all;
```

注意这里权限的设置，否则在复制过程中会报“ORA-01031: insufficient privileges”错误。

（2）由于源端不是本地 UNDO 模式，因此需要以只读模式打开，代码如下：

```
SQL> show pdbs
   CON_ID CON_NAME                       OPEN MODE  RESTRICTED
---------- ------------------------------ ---------- ----------
         2 PDB$SEED                       READ ONLY  NO
         3 ORCLPDB1                       MOUNTED
```

```
SQL> alter session set container=orclpdb1;
Session altered.
SQL> alter pluggable database orclpdb1 open read only;
Pluggable database altered.
SQL> show pdbs
   CON_ID CON_NAME                  OPEN MODE  RESTRICTED
---------- ------------------------------ ---------- ----------
     2 PDB$SEED               READ ONLY  NO
     3 ORCLPDB1               READ ONLY  NO
```

（3）在目标端创建 DBlink。

创建目标 CDB 到源 CDB 之间的 DBlink，首先在 tnsnames.ora 文件中添加如下内容：

```
dave =
  (DESCRIPTION =
   (ADDRESS = (PROTOCOL = TCP)(HOST = 192.168.1.242)(PORT = 1521))
   (CONNECT_DATA =
    (SERVER = DEDICATED)
    (SERVICE_NAME = ORCLCDB) #这里写的是 CDB 的名称
   )
  )
```

接着测试 Service name，代码如下：

```
[oracle@cndba ~]$ tnsping dave
TNS Ping Utility for Linux: Version 18.0.0.0.0 - Production on
24-OCT-2018 20:40:22
Copyright (c) 1997, 2018, Oracle. All rights reserved.
Used parameter files:
/opt/oracle/product/18c/dbhome_1/network/admin/sqlnet.ora
Used TNSNAMES adapter to resolve the alias
Attempting to contact (DESCRIPTION = (ADDRESS = (PROTOCOL = TCP)
(HOST = 192.168.1.242)(PORT = 1521)) (CONNECT_DATA = (SERVER = DEDICATED)
(SERVICE_NAME = ORCLPDB)))
OK (0 msec)
```

然后创建 DBlink，代码如下：

```
SQL> show con_name
CON_NAME
------------------------------
CDB$ROOT
SQL>create public database link dave_pdb connect to c##dave
```

```
identified by oracle using 'dave';
    Database link created.
```

最后测试 DBlink，代码如下：

```
    SQL> select banner_full from v$version@dave_pdb where rownum=1;
    BANNER_FULL
    --------------------------------------------------------------
    Oracle Database 18c Enterprise Edition Release 18.0.0.0.0 -
Production
    Version 18.3.0.0.0
```

（4）执行复制 PDB 的语句，代码如下：

```
    SQL> create pluggable database dave from orclpdb1@dave_pdb
file_name_convert = ('/opt/oracle/oradata/ORCLCDB/ORCLPDB1','/opt/
oracle/oradata/ORCLCDB/dave') ;
    Pluggable database created.
    #dblink 前面的 dave 是远端的 PDB 名称
```

这里如果出现“ORA-1276 File has an Oracle Managed Files file name”错误，可以参考博客文章《12c 创建 PDB 报错 ORA-01276》(https://www.cndba.cn/Expect-le/article/136)，先设置 DB_CREATE_FILE_DEST 参数，再执行 create pluggable database dave from orclpdb1@dave_pdb 命令。具体代码如下：

```
SQL> show pdbs
   CON_ID CON_NAME                  OPEN MODE  RESTRICTED
---------- ------------------------------ ---------- ----------
     2 PDB$SEED                 READ ONLY  NO
     3 ORCLPDB1                 MOUNTED
     4 DAVE                 MOUNTED
SQL> alter pluggable database dave open;
Pluggable database altered.
SQL> show pdbs
   CON_ID CON_NAME                  OPEN MODE  RESTRICTED
---------- ------------------------------ ---------- ----------
     2 PDB$SEED                 READ ONLY  NO
     3 ORCLPDB1                 MOUNTED
     4 DAVE                 READ WRITE NO
```

至此，复制成功。如果是复制 Non-CDB，则最后还要执行脚本 @$ORACLE_HOME/rdbms/admin/noncdb_to_pdb.sql。

在操作过程中可能会遇到如下两个错误。

- ORA-17627: ORA-12514: TNS:listener does not currently know of service requested in connect descriptor

  ORA-17629: Cannot connect to the remote database server

这个错误可以参考笔者的博客文章《Oracle 18c 通过 service_name 连接 PDB 报 ORA-12514 错误的解决方法》(https://www.cndba.cn/dave/article/2997)。

- ORA-65005: missing or invalid file name pattern for file -/opt/oracle/oradata/ORCLCDB/ORCLPDB1/system01.dbf

这个错误与 file_name_convert 参数有关系，可以结合具体的环境，输入 PDB 名称或绝对路径，如“file_name_convert = （'orclpdb1','dave'）;”。

2.5.3 可刷新的 PDB

1. 搭建可刷新的 PDB

可刷新的 PDB 复制远端的 PDB，然后指定一个刷新周期（例如每隔一小时刷新一次数据），这样就能够保持远端 PDB 的数据和目标端 PDB 的数据同步。对于 DataGurad 而言，可刷新的 PDB 的操作对象更小，也更方便。

搭建可刷新 PDB 的准备步骤和复制普通 PDB 的准备步骤相似，这里不再赘述，只是在最后一步创建 PDB 时的语法不同（每隔一小时刷新一次），示例如下：

```
SQL> create pluggable database dave_fresh from orclpdb1@dave_pdb file_name_convert = ('/opt/oracle/oradata/ORCLCDB/ORCLPDB1','/opt/oracle/oradata/ORCLCDB/dave_freash') refresh mode every 1 minutes;
Pluggable database created.
```

可刷新的 PDB 只能以只读模式打开，以读写模式打开会报错，这一点和 DataGurad 类似，代码如下：

```
SQL> alter pluggable database dave_fresh open;
alter pluggable database dave_fresh open
*
ERROR at line 1:
ORA-65341: cannot open pluggable database in read/write mode
SQL> alter pluggable database dave_fresh open read only;
Pluggable database altered.
SQL> show pdbs
  CON_ID CON_NAME                    OPEN MODE  RESTRICTED
```

```
---------- ------------------------------ ---------- ----------
     2 PDB$SEED                READ ONLY  NO
     3 ORCLPDB1                MOUNTED
     6 DAVE            MOUNTED
     7 DAVE_FRESH                  READ ONLY  NO
```

2. 测试数据同步

源端的代码如下：

```
SQL> show pdbs
   CON_ID CON_NAME                   OPEN MODE  RESTRICTED
---------- ------------------------------ ---------- ----------
     2 PDB$SEED                READ ONLY  NO
     3 ORCLPDB1                READ ONLY  NO
SQL> alter pluggable database orclpdb1 close;
Pluggable database altered.
SQL> alter pluggable database orclpdb1 open read write;
Pluggable database altered.
SQL> alter session set container=orclpdb1;
Session altered.
SQL> create table zhixintech (name varchar2 (50)) ;
Table created.
SQL> insert into zhixintech values ('www.cndba.cn') ;
1 row created.
SQL> commit;
Commit complete.
```

目标端的代码如下：

```
SQL> select * from zhixintech;
select * from zhixintech
              *
ERROR at line 1:
ORA-00942: table or view does not exist
```

查询结果是没有数据，我们前面配置的策略是一分钟刷新一次，查看一下日志，代码如下：

```
2018-10-24T23:20:37.378646+08:00
DAVE_FRESH (7) :alter pluggable database refresh
DAVE_FRESH (7) :Completed: alter pluggable database refresh
2018-10-24T23:21:37.303232+08:00
DAVE_FRESH (7) :alter pluggable database refresh
DAVE_FRESH (7) :Completed: alter pluggable database refresh
```

```
2018-10-24T23:22:37.496761+08:00
DAVE_FRESH（7）:alter pluggable database refresh
DAVE_FRESH（7）:Completed: alter pluggable database refresh
```

刷新的动作一直在做，只是没有同步数据，这是因为PDB为只读状态，无法同步数据，需要PDB在挂载状态（Mount）时才能同步，没有类似ADG的open read only with apply 的功能，可以关闭PDB后，再启动PDB，代码如下：

```
SQL> alter pluggable database dave_fresh close;
Pluggable database altered.
SQL> alter session set container=dave_fresh;
Session altered.
SQL> alter pluggable database refresh; #这里是手动刷新
Pluggable database altered.
SQL> alter pluggable database dave_fresh open read only;
Pluggable database altered.
SQL> select * from zhixintech;
NAME
------------------------------------------------
www.cndba.cn
```

数据同步成功。

同步模式也可以进行修改，代码如下：

```
SQL> alter pluggable database dave_fresh refresh mode every 10 minutes; #修改刷新时间
Pluggable database altered.
SQL> alter pluggable database dave_fresh refresh mode manual; #改为手动刷新模式
Pluggable database altered.
SQL> col pdb_name for a20
SQL> select pdb_id, pdb_name, refresh_mode, refresh_interval from dba_pdbs;
    PDB_ID PDB_NAME             REFRESH_MODE REFRESH_INTERVAL
---------- -------------------- ------------ ----------------
         7 DAVE_FRESH           MANUAL
```

转换目标库PDB为不可刷新的PDB，代码如下：

```
SQL> alter pluggable database dave_fresh close;
Pluggable database altered.
SQL> alter pluggable database dave_fresh REFRESH MODE NONE;
Pluggable database altered.
SQL> alter pluggable database dave_fresh REFRESH MODE manual;
```

```
alter pluggable database dave_fresh REFRESH MODE manual
*
ERROR at line 1:
ORA-65261: pluggable database DAVE_FRESH not enabled for refresh
```

注意，禁用 PDB 的可刷新功能后，就不能再转换为可刷新的 PDB 了。

3. 测试 Switchover（转接）

可刷新的 PDB 除能定期同步数据外，还可以进行源端和目标端的转接。我们在刚创建的可刷新的 PDB（dave_fresh）上进行测试。

在源端进行如下配置，在 tnsnames.ora 文件中添加以下代码：

```
dave =
  (DESCRIPTION =
    (ADDRESS = (PROTOCOL = TCP)(HOST = 192.168.1.243)(PORT = 1521))
    (CONNECT_DATA =
      (SERVER = DEDICATED)
      (SERVICE_NAME =ORCLCDB) #这里写的是 CDB 的名称
    )
  )
```

这里必须写 CDB 的名称，否则在转接时会报如下错误：

```
ORA-02019: connection description for remote database not found
```

在源 CDB 中创建指向目标 CDB 的 DBlink，注意，这里的 DBlink 必须是 CDB 级别的，代码如下：

```
SQL> show con_name
CON_NAME
------------------------------
ORCLPDB1
SQL> create public database link dave_pdb connect to c##dave
identified by oracle using 'dave';
Database link created.
```

测试 DBlink，代码如下：

```
SQL> select banner_full from v$version@dave_pdb where rownum=1;
BANNER_FULL
--------------------------------------------------------------
Oracle Database 18c Enterprise Edition Release 18.0.0.0.0 -
Production
Version 18.3.0.0.0
```

进行转接，代码如下：

```
SQL>alter session set container=orclpdb1;
Session altered.
SQL> alter pluggable database refresh mode manual from
DAVE_FRESH@dave_pdb switchover;
alter pluggable database refresh mode manual from
dave_fresh@dave_pdb switchover
*
ERROR at line 1:
ORA-12754: Feature PDB REFRESH SWITCHOVER is disabled due to
missing capability
```

这里报错并显示转接被禁用，原因是该转接功能被限制在 Exadata 上的 Oracle 18c，当前环境为虚拟机，故触发该问题。可通过将隐含参数 _exadata_feature_on 的参数值修改为 true 来解决。

分别在两个数据库的 CDB 上执行以下代码：

```
SQL> alter system set "_exadata_feature_on"=true scope=spfile;
System altered.
SQL> shutdown immediate
SQL> startup
SQL> alter pluggable database orclpdb1 open;
SQL> alter pluggable database refresh mode manual from
DAVE_FRESH@dave_pdb switchover;
Pluggable database altered.
```

转接完成，验证代码如下：

```
SQL> show pdbs
   CON_ID CON_NAME                       OPEN MODE  RESTRICTED
---------- ------------------------------ ---------- ----------
         3 ORCLPDB1                       MOUNTED

SQL> show pdbs

   CON_ID CON_NAME                       OPEN MODE  RESTRICTED
---------- ------------------------------ ---------- ----------
         2 PDB$SEED                       READ ONLY  NO
         3 ORCLPDB1                       MOUNTED
         6 DAVE                           MOUNTED
         7 DAVE_FRESH                     READ WRITE NO
```

2.6 迁移 PDB

迁移（Relocate）PDB 在本质上和复制 PDB 差不多，只是迁移 PDB 是将源端的 PDB 移动到目标端，而复制 PDB 是复制一个一模一样的 PDB。在迁移过程中，源 PDB 保持读写状态并正常提供服务。当第一次打开目标端的 PDB 时，Oracle 将结束源 PDB 上的所有会话，并将客户端连接到新的 PDB 上来。当打开新的 PDB 时，源 PDB 将会自动关闭并从 CDB 中删除，而且只有一个 PDB 保持打开状态。

迁移 PDB 是一种停机时间最短，甚至不用停机的一种移动 PDB 的方法。如果通过插拔数据库的方式移动 PDB，则会大大增加数据库的停机时间，这对于现在 7×24 小时运行的业务来说是绝对不允许的。

2.6.1 迁移 PDB 的前提条件

- 源 CDB 必须是本地 UNDO 模式。
- 如果目标 CDB 是非归档模式，则目标 PDB 只能是只读模式；如果目标 PDB 是归档模式，则没有此限制。
- 源 CDB 和目标 CDB 必须具有相同的字节存储顺序。
- 源 CDB 上已安装的组件要么和目标 CDB 上已安装的组件相同，要么是其子集。
- 如果目标库字符集是 AL32UTF8，则源库字符集可以是任何字符集，这是利用了 Oracle 12c 中引入的一个特性，即当 CDB 字符集是 AL32UTF8 时，同一个 CDB 中不同的 PDB 可以使用不同的字符集。

2.6.2 迁移 PDB 的实现方式

PDB 迁移通过 DBlink 进行在线块级别的复制，或者复制源 PDB 的数据文件（redo 文件和 undo 文件），与此同时，源 PDB 保持着读写模式，对外提供服务。当数据迁移完毕时，只需要打开目标端的 PDB，Oracle 就会断开源 PDB 上的所有会话并关闭源 PDB。

Oracle 18c 引入了更加通用的数据库会话中断机制，利用计时器来断开活动会话。当计时器过期后，Oracle 会中断所有的活动会话，并将这些会话重新连接到迁移后的 PDB 上。在默认情况下，该特性支持数据库服务和 PDB 级别上调用的所有维护操作，如停止服务、迁移服务、迁移 PDB 和停止 PDB。

2.6.3 具体操作示例

在迁移 PDB 之前，要确认源 PDB 为本地 UNDO 模式。

（1）在目标 CDB 中创建管理用户和 DBlink，连接到源 CDB。

创建用户并赋权限，代码如下：

```
SQL> create user c##dave identified by oracle;

SQL> grant sysoper,create session, resource, create any table,
unlimited tablespace ,create pluggable database to c##dave
container=all;
```

目标库配置 TNS，代码如下：

```
[oracle@dave admin]$ cat tnsnames.ora

dave=
  (DESCRIPTION =
    (ADDRESS = (PROTOCOL = TCP)(HOST = 192.168.1.243)(PORT = 1521))
    (CONNECT_DATA =
      (SERVER = DEDICATED)
      (SERVICE_NAME = ORCLCDB)
    )
  )
```

创建 DBlink，代码如下：

```
SQL> show con_name

CON_NAME
------------------------------
CDB$ROOT
SQL> create public database link r_link connect to c##dave
identified by oracle using 'dave';
Database link created.
```

（2）确认源 CDB 和目标 CDB 满足迁移条件。

若源 PDB 是本地 UNDO 模式，则在迁移之前务必再次确认，否则会导致迁移失败。可以通过以下两种方式来确认。

- 通过视图查看，代码如下：

```
SQL> show pdbs
```

```
   CON_ID CON_NAME              OPEN MODE  RESTRICTED
---------- ------------------------------ ---------- ----------
       2 PDB$SEED              READ ONLY  NO
       3 ORCLPDB1              MOUNTED
       6 DAVE              READ WRITE NO
       7 DAVE_FRESH                  READ WRITE NO
SQL> col PROPERTY_NAME for a30;
SQL>col PROPERTY_VALUE for a20;
SQL>select PROPERTY_NAME,PROPERTY_VALUE from database_properties
where property_name='LOCAL_UNDO_ENABLED';

PROPERTY_NAME                 PROPERTY_VALUE
------------------------------ --------------------
LOCAL_UNDO_ENABLED            TRUE
```

- 切换到 PDB 下，查看是否有 UNDO 表空间，代码如下：

```
SQL> alter session set container=dave;
Session altered.
SQL> select name from v$datafile where name like '%undo%';
NAME
--------------------------------------------------------------
/opt/oracle/oradata/ORCLCDB/ORCLCDB/78FD93ED72B869ABE053F301A8
C03231/datafile/o1_mf_undotbs1_fx1b415z_.dbf
```

源 CDB 必须是归档模式，代码如下：

```
SQL> archive log list;
Database log mode            No Archive Mode
Automatic archival           Disabled
Archive destination
/opt/oracle/product/18c/dbhome_1/dbs/arch
Oldest online log sequence     5
Current log sequence         7
SQL> shutdown immediate
Database closed.
Database dismounted.
ORACLE instance shut down.
SQL> startup mount
ORACLE instance started.
Total System Global Area 1593835440 bytes
Fixed Size           8896432 bytes
Variable Size          1023410176 bytes
```

```
Database Buffers  553648128 bytes
Redo Buffers        7880704 bytes
Database mounted.
SQL> alter database archivelog;
Database altered.
SQL> alter database open;
Database altered.
SQL> alter pluggable database dave open;
Pluggable database altered.
SQL> show pdbs

   CON_ID CON_NAME              OPEN MODE  RESTRICTED
---------- ------------------------------ ---------- ----------
     2 PDB$SEED           READ ONLY  NO
     3 ORCLPDB1           MOUNTED
     6 DAVE           READ WRITE NO
     7 DAVE_FRESH               MOUNTED
```

查看源 CDB 的字节码和目标 PDB 的字节码是否相同，这里的查询结果都是 Little，代码如下：

```
SQL> set lines 120
SQL> col platform_name for a20
SQL> select
    db.name,
    db.platform_id,
    db.platform_name ,
    os.ENDIAN_FORMAT
  from
    v$database db ,v$transportable_platform os
  where db.platform_id=os.platform_id;
NAME      PLATFORM_ID PLATFORM_NAME    ENDIAN_FORMAT
------------------ --------------------------------------------
ORCLCDB         13 Linux x86 64-bit    Little
```

查看源 CDB 的字符集和目标 CDB 的字符集，这里都是 AL32UTF8，代码如下：

```
SQL> col PROPERTY_NAME for a30
SQL> col PROPERTY_VALUE for a40
SQL>SELECT PROPERTY_NAME, PROPERTY_VALUE
FROM DATABASE_PROPERTIES
WHERE PROPERTY_NAME ='NLS_CHARACTERSET';
PROPERTY_NAME             PROPERTY_VALUE
```

```
-----------------------------------------------------------------
NLS_CHARACTERSET       AL32UTF8
```

当 CDB 的字符集是 AL32UTF8 时，PDB 可以是任意的字符集。

（3）通过 DBlink 迁移 PDB。

在目标库上执行以下命令，PDB 的大小和带宽不同，创建的结果也会有所不同，代码如下：

```
    SQL> create pluggable database dave from dave@r_link relocate
availability normal;
    create pluggable database dave from dave@r_link relocate
availability normal
                                                                    *
    ERROR at line 1:
    ORA-65016: FILE_NAME_CONVERT must be specified
    SQL> alter system set
DB_CREATE_FILE_DEST='/opt/oracle/oradata/ORCLCDB';
    System altered.
    SQL> create pluggable database dave from dave@r_link relocate
availability normal;
    Pluggable database created.
```

在源 PDB 上执行创建表并插入数据的操作，以验证数据是否同步，代码如下：

```
SQL> show pdbs
   CON_ID CON_NAME                 OPEN MODE  RESTRICTED
---------- ------------------------------ ---------- ----------
     2 PDB$SEED              READ ONLY  NO
     3 ORCLPDB1              MOUNTED
     6 DAVE            READ WRITE NO
     7 DAVE_FRESH                 MOUNTED
SQL> alter session set container=dave;
Session altered.
SQL> create table ahdba as select * from all_objects;
Table created.
```

启动新创建的目标 PDB，并查看新建的表和数据是否存在，代码如下：

```
SQL> show pdbs
   CON_ID CON_NAME                 OPEN MODE  RESTRICTED
---------- ------------------------------ ---------- ----------
     2 PDB$SEED              READ ONLY  NO
     3 ORCLPDB1              MOUNTED
```

```
      5 DAVE            MOUNTED
SQL> alter pluggable database dave open;
Pluggable database altered.
SQL> alter session set container=dave;
Session altered.
SQL> select count (1) from ahdba;

  COUNT (1)
----------
     71883
```

当目标 PDB 被打开后，源 PDB 将会从源 CDB 中自动删除，代码如下：

```
SQL> show pdbs
   CON_ID CON_NAME                       OPEN MODE  RESTRICTED
---------- ------------------------------ ---------- ----------
         2 PDB$SEED                       READ ONLY  NO
         3 ORCLPDB1                       MOUNTED
         7 DAVE_FRESH                     MOUNTED
```

我们可以查看源库的日志，里面详细地记录了迁移 PDB 的过程，代码如下：

```
2018-10-25T10:31:40.557133+08:00
DAVE (6) :Process termination requested for pid 29806 [source = rdbms], [info = 2] [request issued by pid: 31788, uid: 54321]
2018-10-25T10:31:40.606875+08:00
DAVE (6) :KILL SESSION for sid= (1, 7562) :
DAVE (6) :  Reason = PDB close immediate
DAVE (6) :  Mode = KILL HARD FORCE -/-/-
DAVE (6) :  Requestor = USER (orapid = 58, ospid = 31788, inst = 1)
DAVE (6) :  Owner = Process: USER (orapid = 33, ospid = 29806)
DAVE (6) :  Result = ORA-0
2018-10-25T10:31:41.606672+08:00
Pluggable database DAVE closed
DAVE (6) :JIT: pid 31788 requesting stop
DAVE (6) :Buffer Cache flush started: 6
DAVE (6) :Buffer Cache flush finished: 6
2018-10-25T10:31:41.903417+08:00
DAVE (6) :While transitioning the pdb 6 to clean state, clearing all its abort bits in the control file.
Pluggable database DAVE closed
2018-10-25T10:31:45.474492+08:00
Deleted Oracle managed file
```

```
/opt/oracle/oradata/ORCLCDB/ORCLCDB/78FD93ED72B869ABE053F301A8C032
31/datafile/o1_mf_users_fx1b4160_.dbf
    Deleted Oracle managed file
/opt/oracle/oradata/ORCLCDB/ORCLCDB/78FD93ED72B869ABE053F301A8C032
31/datafile/o1_mf_temp_fx1b4160_.dbf
    Deleted Oracle managed file
/opt/oracle/oradata/ORCLCDB/ORCLCDB/78FD93ED72B869ABE053F301A8C032
31/datafile/o1_mf_undotbs1_fx1b415z_.dbf
    Deleted Oracle managed file
/opt/oracle/oradata/ORCLCDB/ORCLCDB/78FD93ED72B869ABE053F301A8C032
31/datafile/o1_mf_sysaux_fx1b415z_.dbf
    Deleted Oracle managed file
/opt/oracle/oradata/ORCLCDB/ORCLCDB/78FD93ED72B869ABE053F301A8C032
31/datafile/o1_mf_system_fx1b415y_.dbf
```

至此，迁移 PDB 完成。如果是在生产环境中，则建议对新创建的 PDB 进行一次备份。

2.7　插入 PDB

在插入 PDB 时有以下两个因素需要考虑。

- 源 CDB 和目标 CDB 要有相同的字节存储顺序。
- 字符集要么相同，要么是子集，和之前复制部分的要求一样。

操作时在 CREATE PLUGGABLE DATABASE 中使用 USING 子句，指定一个 XML 的 PDB 元数据文件或一个压缩的 PDB 归档文件（.pdb 类型的文件）来创建一个 PDB。

XML 元数据文件描述了未插入的 PDB 和与 PDB 相关的文件（例如数据文件和钱包文件）。归档文件包括 XML 元数据文件和 PDB 文件。当指定 XML 元数据文件时，XML 文件包含 PDB 文件的完整路径。当指定.pdb 归档文件时，XML 元数据文件只包含相对文件名。通过.pdb 归档文件或 XML 元数据文件，可以将 PDB 插到 CDB 中或应用程序 root 中。

2.7.1　数据文件存储目录

通过.pdb 归档文件插入 PDB，不需要添加参数来指定数据文件的存放位置，因为 Oracle 会将归档文件解压缩到当前目录，当前目录就是数据文件的存放目录。

通过 XML 元数据文件插入 PDB，需要添加 SOURCE_FILE_NAME_CONVERT 或 SOURCE_FILE_DIRECTORY 子句来指定文件存放路径。因为 XML 元数据文件里存放的是 PDB 之前的数据文件的绝对路径，该路径不一定存在，所以需要指定该参数。

2.7.2 操作示例：将 PDB 拔出并插到另一个 CDB 中

本示例将之前迁移出来的 PDB dave 再迁移回去。

1. 拔出 PDB

拔出 PDB 的代码如下：

```
SQL> show pdbs
   CON_ID CON_NAME        OPEN MODE  RESTRICTED
---------- ------------------------------ ---------- ----------
     2 PDB$SEED              READ ONLY  NO
     3 ORCLPDB1              MOUNTED
     5 DAVE             READ WRITE NO
SQL> alter session set container=dave;
Session altered.
```

拔出 PDB 有两种方式：XML 方式和归档文件方式。

（1）以 XML 方式将源 PDB 从 CDB 中拔出

查询 PDB 的数据文件存放路径，代码如下：

```
SQL> select name from v$datafile;

NAME
--------------------------------------------------------------
/opt/oracle/oradata/ORCLCDB/ORCLCDB/78FD93ED72B869ABE053F301A8
C03231/datafile/o1_mf_system_fx29v6y4_.dbf
/opt/oracle/oradata/ORCLCDB/ORCLCDB/78FD93ED72B869ABE053F301A8
C03231/datafile/o1_mf_sysaux_fx29v6y9_.dbf
/opt/oracle/oradata/ORCLCDB/ORCLCDB/78FD93ED72B869ABE053F301A8
C03231/datafile/o1_mf_undotbs1_fx29v6yb_.dbf
/opt/oracle/oradata/ORCLCDB/ORCLCDB/78FD93ED72B869ABE053F301A8
C03231/datafile/o1_mf_users_fx29v6yc_.dbf
```

关闭需要拔出的 PDB，代码如下：

```
SQL> alter pluggable database dave close;
```

```
Pluggable database altered.
```

将名为 dave 的 PDB 拔出，并生成名为 cndbapdb.xml 的 XML 元数据文件，位于/tmp/目录下，代码如下：

```
SQL> show con_name
CON_NAME
------------------------------
CDB$ROOT
SQL> alter pluggable database dave unplug into '/tmp/dave.xml';
Pluggable database altered.
SQL> show pdbs

   CON_ID CON_NAME                       OPEN MODE  RESTRICTED
---------- ------------------------------ ---------- ----------
         2 PDB$SEED                       READ ONLY  NO
         3 ORCLPDB1                       MOUNTED
         5 DAVE                       MOUNTED
```

（2）以归档文件方式将源 PDB 从 CDB 中拔出

如果使用归档文件方式拔出 PDB，则所有数据文件和 XML 元数据文件将被压缩到一个归档文件中（以.pdb 结尾）。归档文件可以节省空间和网络传输时间，只需要复制一个文件即可，代码如下：

```
SQL> alter pluggable database dave unplug into '/tmp/dave. pdb'  ;
```

2. 复制相关数据文件到目标端

在目标端创建必要的目录，并将源端的相关数据文件复制到目标端。如果是归档文件，则复制一个归档文件即可，代码如下：

```
[oracle@dave admin]$ cd
/opt/oracle/oradata/ORCLCDB/ORCLCDB/78FD93ED72B869ABE053F301A8C032
31/datafile/
[oracle@dave datafile]$ ls
o1_mf_sysaux_fx29v6y9_.dbf  o1_mf_system_fx29v6y4_.dbf
o1_mf_temp_fx29v6yc_.dbf  o1_mf_undotbs1_fx29v6yb_.dbf
o1_mf_users_fx29v6yc_.dbf
[oracle@dave datafile]$ scp *
192.168.1.243:/opt/oracle/oradata/ORCLCDB/dave
oracle@192.168.1.243's password:
o1_mf_sysaux_fx29v6y9_.dbf          100%  390MB  64.9MB/s   00:06
o1_mf_system_fx29v6y4_.dbf          100%  300MB  60.0MB/s   00:05
o1_mf_temp_fx29v6yc_.dbf            100%   62MB  62.0MB/s   00:01
```

```
o1_mf_undotbs1_fx29v6yb_.dbf     100%  100MB  32.4MB/s   00:03
o1_mf_users_fx29v6yc_.dbf        100% 5128KB  24.2MB/s   00:00
[oracle@dave datafile]$ scp /tmp/dave.xml 192.168.1.243:/opt/
oracle/oradata/ORCLCDB/dave
oracle@192.168.1.243's password:
dave.xml                         100% 8176     6.9MB/s   00:00
[oracle@dave datafile]$
```

3. 检查源 PDB 和目标 CDB 是否兼容

如果检查结果是 YES，则表示兼容，可以正常插到目标 CDB 中；如果检查结果是 NO，则可以打开 PDB_PLUG_IN_VIOLATIONS 视图来查看为什么不兼容，代码如下：

```
SQL> SET SERVEROUTPUT ON
SQL> DECLARE
  compatible CONSTANT VARCHAR2（3） :=
    CASE DBMS_PDB.CHECK_PLUG_COMPATIBILITY（
          pdb_descr_file => '/opt/oracle/oradata/ORCLCDB/dave/
dave.xml',
          pdb_name      => 'dave'）
    WHEN TRUE THEN 'YES'
    ELSE 'NO'
END;
BEGIN
  DBMS_OUTPUT.PUT_LINE（compatible）;
END;
/
NO

PL/SQL procedure successfully completed.
```

若在 Oracle 18c 中执行时出现 bug，则可以参考笔者的博客文章《Oracle 18c bug 执行 DBMS_PDB.CHECK_PLUG_COMPATIBILITY 报 ORA-7445 的解决方法》（https://www.cndba.cn/dave/article/3096）。

本示例检查结果返回的是 NO，查看视图，代码如下：

```
SQL> select message from PDB_PLUG_IN_VIOLATIONS where name='DAVE';
MESSAGE
---------------------------------------------------------------
'18.3.0.0.0 Release_Update 1806280943' is installed in the CDB but
'18.3.0.0.0 R
elease_Update 180628094' is installed in the PDB
```

我们刚打的补丁导致两边的版本不一样，所以检测没有通过。我们这里忽略这个错误，继续操作。

4. 插到目标 CDB 中

如果源端和目标端的数据文件的路径不同，则需要使用 source_file_name_convert 命令，代码如下：

```
    SQL> create pluggable database dave using
'/opt/oracle/oradata/ORCLCDB/dave/dave.xml' source_file_name_convert =
('/opt/oracle/oradata/ORCLCDB/ORCLCDB/78FD93ED72B869ABE053F301A8C03231/
datafile/','/opt/oracle/oradata/ORCLCDB/dave/') nocopy tempfile reuse;
    Pluggable database created.
    SQL> show pdbs
       CON_ID CON_NAME                   OPEN MODE  RESTRICTED
    ---------- ------------------------------ ---------- ----------
         2 PDB$SEED              READ ONLY  NO
         3 ORCLPDB1              MOUNTED
         4 DAVE             MOUNTED
         7 DAVE_FRESH                 MOUNTED
    SQL> alter pluggable database dave open;
    Pluggable database altered.
    SQL> show pdbs
       CON_ID CON_NAME                   OPEN MODE  RESTRICTED
    ---------- ------------------------------ ---------- ----------
         2 PDB$SEED              READ ONLY  NO
         3 ORCLPDB1              MOUNTED
         4 DAVE             READ WRITE NO
         7 DAVE_FRESH                 MOUNTED
    SQL> alter session set container=dave;
    Session altered.
    SQL> select name from v$datafile;
    NAME
    --------------------------------------------------------------
    /opt/oracle/oradata/ORCLCDB/dave/o1_mf_system_fx29v6y4_.dbf
    /opt/oracle/oradata/ORCLCDB/dave/o1_mf_sysaux_fx29v6y9_.dbf
    /opt/oracle/oradata/ORCLCDB/dave/o1_mf_undotbs1_fx29v6yb_.dbf
    /opt/oracle/oradata/ORCLCDB/dave/o1_mf_users_fx29v6yc_.dbf
```

如果源端和目标端的数据文件的路径相同，则使用以下语句:

```
    SQL>create pluggable database dave using '/opt/oracle/oradata/
orclcdb/dave/dave.xml' nocopy tempfile reuse;
```

如果通过.pdb 归档文件来插入 PDB，则需要修改以下文件名：

```
SQL>create pluggable database dave using '/u01/app/oracle/
oradata/dave/dave.pdb' nocopy tempfile reuse;
```

5. 删除源 PDB

迁移完成后，可以直接删除源库。如果不需要数据文件，则要在语句中加上“including datafiles”，代码如下：

```
SQL> show pdbs
   CON_ID CON_NAME                  OPEN MODE  RESTRICTED
---------- ------------------------------ ---------- ----------
     2 PDB$SEED             READ ONLY  NO
     3 ORCLPDB1             MOUNTED
     5 DAVE             MOUNTED
SQL> drop pluggable database dave including datafiles;
Pluggable database dropped.
SQL> show pdbs
   CON_ID CON_NAME                  OPEN MODE  RESTRICTED
---------- ------------------------------ ---------- ----------
     2 PDB$SEED             READ ONLY  NO
     3 ORCLPDB1             MOUNTED
```

2.8　移除 PDB

前面讲解了 CDB 和 PDB 的创建，移除 PDB 相对于创建 PDB 来说简单很多，可以通过拔出 PDB、删除 PDB、迁移 PDB 三种方式把 CDB 中的 PDB 移除。拔出和迁移 PDB 在前面已经讲过，这里不再重复，下面简单介绍一下删除 PDB。

删除 PDB 不会删除之前 PDB 的所有备份和归档日志文件，这些需要使用 RMAN 命令来删除。删除 PDB 有两种方式：一种是删除 PDB，但是保留所有数据文件；另一种是将与 PDB 相关的所有数据文件全部删除。

注意下面的两个语法。

- KEEP DATAFILES：即使使用 KEEP DATAFILES 命令，临时文件也会被删除。

```
SQL> DROP PLUGGABLE DATABASE cndbapdb KEEP DATAFILES;
```

- INCLUDING DATAFILES：如果该 PDB 是通过 SNAPSHOT COPY 方式创建的，则必须采用这种方式来删除。

```
SQL> DROP PLUGGABLE DATABASE cndbapdb INCLUDING DATAFILES;
```

第3章

管理多租户环境

3.1　CDB 字符集

从 Oracle 12.2 开始，PDB 可以设置和 CDB 不同的字符集，也就是插入 CDB 或复制的 PDB 的字符集可以和 CDB root 的字符集不同，但前提是 CDB 字符集必须是 AL32UTF8（通过 DBCA 创建数据库时默认也是该字符集）。当 CDB root 的字符集不是 AL32UTF8 时，CDB 中的所有 PDB 都必须和 CDB root 的字符集相同。

Oracle 官方建议 CDB root 字符集使用 AL32UTF8、CDB 国家字符集使用 AL16UTF16，这样就拥有了最大程度的灵活性，不管是什么字符集的 PDB，都可以移动或迁移到该 CDB 中。

表 3-1 是不同版本的 Oracle 支持的 Unicode 字符集及 Unicode 版本。

表 3-1　不同版本的 Oracle 支持的 Unicode 字符集及 Unicode 版本

字符集	支持的数据库版本	Unicode 编码格式	Unicode 版本	数据库字符集	国家字符集
AL24UTFFSS	7.2~8i	UTF-8	1.1	是	不是
UTF8	8.0~12c	CESU-8	Oracle 8.0~8.1.6：2.1 Oracle 8.1.7 以后：3.0	是	是 （Oracle 9i 以后）
UTFE	8.0~12c	UTF-EBCDIC	Oracle 8.0~8.1.6：2.1 Oracle8.1.7 以后：3.0	是	不是

续表

字符集	支持的数据库版本	Unicode编码格式	Unicode 版本	数据库字符集	国家字符集
AL32UTF8	9i 到 18	UTF-8	Oracle 9iR1 1：3.0 Oracle 9iR2 2：3.1 Oracle 10gR1：3.2 Oracle 10gR2：4.0 Oracle 11g：5.0 Oracle 12cR1：6.2 Oracle 12cR2 2：7.0 Oracle 18cR1：9.0	是	不是
AL16UTF16	9i 到 18	UTF-16	Oracle 9iR1：3.0 Oracle 9iR2：3.1 Oracle 10gR1：3.2 Oracle 10gR2：4.0 Oracle 11g：5.0 Oracle 12cR1：6.2 Oracle 12cR2：7.0 Oracle 18cR1：9.0	不是	是

3.2 管理 CDB

管理 CDB 和管理 Non-CDB 相似，但也有一些不同的地方，主要体现在管理任务上，有些是应用到整个 CDB 上，而有些只能应用到 PDB 上。

3.2.1 当前容器

CDB 中的每个容器的数据字典都是独立的。当前容器可以是 CDB root、应用程序 root、PDB 或应用程序 PDB。每个会话同时只能连接一个容器，但会话可以从一个容器切换到另一个容器。

如果当前容器是 CDB root，那么只能用公共用户来连接；如果当前容器是 PDB，那么可以指定公共用户和本地用户来连接；如果 SQL 语句中包含 CONTAINER=ALL，那么当前容器必须是 CDB root 或应用程序 root。

每个容器都有唯一的 CON_ID 和 CON_NAME，可以用以下两种方式来查看当前容器的 CON_ID 和 CON_NAME。

- 通过 SYS_CONTEXT 命令来查看，代码如下：

```
SQL> COL CON_ID FORMAT A10
SQL>COL CON_NAME FORMAT A20
SQL>SELECT SYS_CONTEXT ('USERENV', 'CON_ID') AS
CON_ID,SYS_CONTEXT ('USERENV', 'CON_NAME') AS CON_NAME FROM DUAL;
CON_ID     CON_NAME
---------- --------------------
1      CDB$ROOT
```

- 通过 show 命令来查看，代码如下：

```
SQL> show con_name
CON_NAME
------------------------------
CDB$ROOT
SQL> show con_id
CON_ID
------------------------------
1
```

使用 alter session 命令可以切换当前容器，代码如下：

```
SQL> show pdbs
   CON_ID CON_NAME          OPEN MODE  RESTRICTED
---------- ------------------------------ ---------- ----------
     2 PDB$SEED            READ ONLY  NO
     3 ORCLPDB1            MOUNTED
SQL> alter session set container=orclpdb1;
Session altered.
SQL> show con_name
CON_NAME
------------------------------
ORCLPDB1
```

使用 SQLPLUS 或 CONNECT 命令可以连接指定的 PDB，这种方法需要先配置好 tnsnames.ora 文件，代码如下：

```
[oracle@cndba admin]$ cat tnsnames.ora
# tnsnames.ora Network Configuration File:
/opt/oracle/product/18c/dbhome_1/network/admin/tnsnames.ora
# Generated by Oracle configuration tools.
davepdb =
  (DESCRIPTION =
```

```
    (ADDRESS = (PROTOCOL = TCP)(HOST = 192.168.1.243)(PORT = 1521))
    (CONNECT_DATA =
      (SERVER = DEDICATED)
      (SERVICE_NAME = DAVE)
    )
  )
[oracle@cndba admin]$ sqlplus c##dave/oracle@davepdb
SQL*Plus: Release 18.0.0.0.0 - Production on Thu Oct 25 15:00:27
2018
Version 18.3.0.0.0
Copyright (c) 1982, 2018, Oracle.  All rights reserved.
Last Successful login time: Thu Oct 25 2018 10:31:43 +08:00
Connected to:
Oracle Database 18c Enterprise Edition Release 18.0.0.0.0 -
Production
Version 18.3.0.0.0
SQL> show con_name

CON_NAME
------------------------------
DAVE
SQL> conn c##dave/oracle@davepdb
Connected.
```

3.2.2　CDB 中的管理任务

只有公共用户才能执行 CDB 中的管理任务，当然，公共用户也需要具有相应的权限。如表 3-2 所示是一些常见的 CDB 中的管理任务及其说明。

表 3-2　常见的 CDB 中的管理任务及其说明

任　务	说　明
启动 CDB 实例	必须是公共用户连接到 CDB root，然后执行 startup 命令
进程	CDB 和其内所有容器公用同一组后台进程
内存	CDB 中只有一个 SGA 和一个 PGA。CDB 所需的内存是 CDB 中其他容器（PDB）所需内存的总和
告警日志	整个 CDB 共用一个告警日志
控制文件	整个 CDB 共用一个控制文件
在线 redo 日志和归档日志	整个 CDB 共用一个在线 redo 日志和一组归档日志文件
表空间	可以为 CDB 和每个 PDB 创建、删除、修改表空间（临时）

续表

任　务	说　明
数据文件和临时文件	CDB 和 PDB 有各自的数据文件，管理方式与 Non-CDB 的管理方式有以下两点不同： （1）可以限制每个 PDB 可使用的空间大小（指定存储空间大小）； （2）可以为 CDB 和每个 PDB 创建默认的临时表空间
UNDO 模式	CDB 有两种 UNDO 模式： （1）Local UNDO：CDB 和每个 PDB 都有自己的 UNDO 表空间； （2）Share UNDO：CDB 和 PDB 公用一个 UNDO 表空间。 在 CDB 中，UNDO_MANAGEMENT 初始化参数必须被设置为 AUTO
在容器之间迁移数据	和在 Non-CDB 之间迁移数据一样，可以通过数据泵、RMAN、导出和导入等方式来迁移数据
搭建 DataGuard 备库	可以搭建物理备库、逻辑备库，但是只能在 CDB 级别搭建 DataGuard 备库，无法对单个 PDB 搭建 DataGuard 备库
删除数据库	删除 CDB 时会删除该 CDB 中的所有 PDB 相关文件，也可以指定删除某个容器，如 PDB、应用程序 PDB

3.2.3　修改 CDB 参数

PDB 中的部分参数值继承自 CDB 中的参数值，也可以单独修改。PDB 中每个初始化参数都有一个继承属性（TRUE 或 FALSE）。当 PDB 的参数值继承自 CDB root 的参数值时，继承属性为 TRUE；否则为 FALSE。如果为 TRUE，那么 PDB 会继承 CDB 中该参数的值，即如果修改了 CDB 中该参数的值，那么 PDB 中该参数的值也会被修改；反之，若为 FALSE，则修改 CBD 中该参数的值，不会修改 PDB 中该参数的值。

PDB 中的部分参数的继承属性必须为 TRUE。对于其他参数，当当前容器是 PDB 时，可以通过执行 ALTER SYSTEM SET 语句将继承属性修改为 FALSE。如果想把该参数的继承属性修改回来，可以通过执行 ALTER SYSTEM RESET 语句来修改。

如果 V$SYSTEM_PARAMETER 视图中参数的 ISPDB_MODIFIABLE 列是 TRUE，那么该参数的继承属性是 FALSE。在 Oracle 18c 中，这样的参数一共有 210 个。当当前容器为 CDB root 时，ALTER SYSTEM SET 语句中的 CONTAINER 参数将决定修改哪个 PDB 的参数值。

CONTAINER = { CURRENT | ALL }，CONTAINER 参数有以下两个值。

- CURRENT：只修改当前容器的参数值，如果当前容器是 root，那么也会修改继承属性是 TRUE 的参数值。
- ALL：适用于 CDB 中的所有容器，包括 root 和所有 PDB。

例如，要修改所有容器的参数 open_ cursors，先查看该参数的继承属性是否为 FALSE，代码如下：

```
SQL> col name for a20
SQL>select con_id,name,ispdb_modifiable from v$system_parameter
where name='open_cursors';
    CON_ID NAME         ISPDB_MODI
---------- -------------------- ----------
         0 open_cursors  TRUE
         3 open_cursors  TRUE
```

再查看当前参数值，代码如下：

```
SQL> set lines 120
SQL> show parameter open_cursors

NAME                     TYPE        VALUE
--------------------------- -----------------------
------------------------------
open_cursors                 integer        202
```

如果在 PDB 中修改参数值，那么继承属性就会被修改为 FALSE，该参数值将不会再受 CDB 的影响。

修改当前 root 的参数 open_ cursors，看 PDB 的参数是否也会被修改，步骤如下。

（1）修改 CDB，代码如下：

```
SQL> alter system set open_cursors = 210;
System altered.
SQL> show parameter open_cursors;
NAME                         TYPE         VALUE
-------------------------------------------------------------------
open_cursors                 integer        210
```

（2）查看 PDB，可以看到 PDB 的参数没有受到影响，代码如下：

```
SQL> alter session set container=dave;
Session altered.
```

```
SQL> show parameter open_cursors;
NAME                 TYPE       VALUE
-----------------------------------------------------
open_cursors         integer    210
```

3.2.4　修改 PDB 参数

以公共用户身份连接到 CDB root，可以通过 ALTER PLUGGALE DATABASE 语句来修改 PDB 的相关配置，也可以直接连接到 PDB 中，通过 ALTER DATABASE 语句来修改相关配置。

1．使用 ALTER DATABASE 命令修改 CDB 的参数

使用 ALTER DATABASE 语句可以修改整个 CDB 的参数，包括 root 的参数和其中所有的 PDB 的参数，但并不是所有参数都支持在 CDB 级别中修改，部分参数只能修改 root，而不会修改 PDB。

（1）修改 CDB。当公共用户连接的是 CDB root 时，通过 ALTER DATABASE 命令执行如下语句会修改整个 CDB：

```
    startup/recovery/logfile/controlfile/standbydatabase/instance/
security/RENAME/GLOBAL_NAME/ ENABLE LOCK CHANGE TRACKING /DISABLE LOCK
CHANGE TRACKING
```

（2）只修改 CDB root。当公共用户连接的是 CDB root 时，通过 ALTER DATABASE 命令执行如下语句只修改 CDB root：

```
    datafile / DEFAULT EDITION / DEFAULT TABLESPACE / DEFAULT TEMPORARY
TABLESPACE
```

下面的语句会修改 CDB root 并提供给 PDB 一个默认值：

```
    flashback/SET DEFAULT{BIGFILE|SMALLFILE} TALESPACE/set_time_zone
```

（3）修改一个或多个 PDB。当公共用户连接到 CDB root 时，可以通过 ALTER PLUGGABLE DATABASE 语句修改 PDB 的打开状态（MOUNT/READ ONLY/READ WRITE 等），以及保存/忽略 PDB 的打开状态。

使用 ALTER DATABASE 命令添加、修改、删除数据文件的操作，只针对当前容器，不针对所有容器。而使用 ALTER DATABASE 命令备份控制文件、启动数据库、修改 redo 日志会对整个 CDB 产生影响，包括 CDB root 和所有 PDB。

2. 使用 ALTER DATABASE 命令修改 CDB 的 UNDO 模式

默认情况下是本地 UNDO 模式，每个 PDB 都有自己的 UNDO 表空间。在 Oracle RAC 中，每个节点中的每个 PDB 都有独立的 UNDO 表空间。本地 UNDO 模式使每个 PDB 更加独立，提高了操作效率，例如插入 PDB 或 PDB 基于时间点的恢复。另外，有些操作必须在本地 UNDO 模式下执行，例如迁移 PDB 和复制 PDB。

（1）查看当前 UNDO 模式，代码如下：

```
SQL> col PROPERTY_NAME for a30;
SQL> col PROPERTY_VALUE for a20;
SQL> select PROPERTY_NAME,PROPERTY_VALUE from
database_properties where property_name='LOCAL_UNDO_ENABLED';
PROPERTY_NAME                 PROPERTY_VALUE
------------------------------ --------------------
LOCAL_UNDO_ENABLED            TRUE
```

TRUE 代表本地 UNDO 模式，FLASE 代表共享 UNDO 模式。

（2）共享 UNDO 模式转换为本地 UNDO 模式的步骤如下。

关闭 CDB，代码如下：

```
SQL> SHUTDOWN IMMEDIATE;
```

以 UPGRADE 方式启动 CDB，代码如下：

```
SQL> STARTUP UPGRADE;
```

确认当前容器是 CDB root ，代码如下：

```
SQL> SHOW CON_NAME
CON_NAME
------------------------------
CDB$ROOT
```

启用本地 UNDO，代码如下：

```
SQL> ALTER DATABASE LOCAL UNDO ON;
```

重启 CDB，正常打开，代码如下：

```
SQL> SHUTDOWN IMMEDIATE;
SQL> STARTUP;
```

（3）本地 UNDO 模式转换为共享 UNDO 模式的步骤如下。

关闭 CDB，代码如下：

```
SQL> SHUTDOWN IMMEDIATE;
```

以 UPGRADE 方式启动 CDB，代码如下：

```
SQL> STARTUP UPGRADE;
```

确认当前容器是 CDB root ，代码如下：

```
SQL> SHOW CON_NAME
CON_NAME
------------------------------
CDB$ROOT
```

禁用本地 UNDO，代码如下：

```
SQL> ALTER DATABASE LOCAL UNDO OFF;
```

重启 CDB，正常打开，代码如下：

```
SQL> SHUTDOWN IMMEDIATE;
SQL> STARTUP;
```

3.2.5　CDB 和 PDB 参数保存位置说明

CDB 中的参数文件和 Oracle 12c 版本之前的参数文件一样，使用 SPFILE，而 PDB 的参数有一些变化，它并没有具体的参数文件。在前面的章节中提到 PDB 会继承 CDB 的参数，也可以修改 PDB 的参数，使其与 CDB 的参数不同，这些不同的 PDB 参数会保存在 CDB 的 PDB_SPFILE$字典表中，并以 con_id 区分。当拔出 PDB 时，PDB 参数会被写入 PDB 的 XML 文件中；当执行 drop pluggable database 命令后，PDB 信息和 PDB_SPFILE$记录会被清除。

```
#CDB:
SQL> show con_name
CON_NAME
------------------------------
CDB$ROOT
SQL> show parameter spfile
NAME  TYPE         VALUE
------- -------- ------------------------------
spfile    string
/u01/app/oracle/product/18.3.0/db_1/dbs/spfilecndba.ora
```

```
#PDB:
SQL> alter session set container=dave;
Session altered.
SQL> show parameter spfile
NAME  TYPE         VALUE
-------- --------- ------------------------------
spfile    string
/u01/app/oracle/product/18.3.0/db_1/dbs/spfilecndba.ora
```

在 3.2.3 小节中，我们修改了 PDB 的 open_cursors 参数，这里验证一下：

```
SQL> alter session set container=cdb$root;
Session altered.
SQL> select pdb_uid,name,value$ from PDB_SPFILE$;
PDB_UID   NAME         VALUE$
--------------- ------------------------ -------------
2977624433   open_cursors    210
SQL> select con_id,dbid,con_uid,guid from v$pdbs;
   CON_ID    DBID    CON_UID GUID
---------- ---------- --------------------------------
     2 3078093964 3078093964 7452CF54594D67D9E053A838A8C07A01
     3 2977624433 2977624433 7452EF41762778F1E053A838A8C0BA10
```

3.3 CDB Fleet 特性

Oracle 从 18c 版本开始支持 CDB Fleet，简单地说，Fleet 就是一个逻辑的 CDB 集合。其中，Lead CDB 是 CDB Fleet 中用于监控和管理其他 CDB 的。

将一个 CDB 中的 LEAD_CDB 参数设置为 TRUE 表示该 CDB 是 Lead CDB，CDB Fleet 中的其他 CDB 将配置 LEAD_CDB_URI 参数并指向 Lead CDB。所有 CDB 中的 PDB 信息都会和 Lead CDB 保持同步，所有 PDB 都会在 Lead CDB 中“可见”，也就是 Lead CDB 可以访问任何 PDB 中的信息。CDB Fleet 支持所有 Oracle 特性，例如 RAC、RMAN、基于时间点恢复、闪回等。

在 CDB Fleet 中，可以通过以下方式来访问 CDB 和 PDB 中的数据：

- CDB 视图；
- GV$视图；
- CONTAINERS 子句；
- 容器映射。

这里需要注意以下两点：

- 在 CDB Fleet 中，各个 CDB 中的 PDB 名称不能相同；
- 可以在任意一个 CDB 中创建 PDB，但是只能在创建时所在的 CDB 中打开该 PDB。

CDB Fleet 的适用场景如下：

- 如果要为一个应用创建更多的 PDB（超过 4096 个），则需要创建多个 CDB；
- 在同一个 CDB Fleet 中，可配置不同类型的服务器来运行不同的业务；
- 使用相同应用程序的不同 PDB 必须存储在不同的位置。

3.3.1　配置 CDB Fleet 环境

配置 CDB Fleet 环境十分简单，包括设置 Lead CDB 和分配其他 CDB Fleet 成员。

1．设置 Lead CDB

在相应的 CDB 中将参数 LEAD_CDB 设置为 TRUE 即可启用 Lead CDB，代码如下：

```
SQL> show con_name

CON_NAME
------------------------------
CDB$ROOT
SQL> select property_value  from  database_properties where
property_name='LEAD_CDB';
no rows selected
```

这里没有启用 Lead CDB，通过以下命令可以启动该特性：

```
SQL> alter database set lead_cdb = true;
Database altered.
SQL> select property_value  from  database_properties where
property_name='LEAD_CDB';
PROPERTY_VALUE
--------------------------------------------------
TRUE
```

2. 分配其他 CDB Fleet 成员

将其他 CDB 中的参数 LEAD_CDB_URI 设置为指向 Lead CDB 的 DBlink 即可。DBlink 必须是固定的用户定义，即用户名和密码都必须出现在 DBlink 的语法中。在其他 CDB 中进行如下配置。

配置 tnsnames.ora 文件，代码如下：

```
[oracle@dave admin]$ cat tnsnames.ora
dave=
  (DESCRIPTION =
    (ADDRESS = (PROTOCOL = TCP)(HOST = 192.168.1.243)(PORT = 1521))
    (CONNECT_DATA =
      (SERVER = DEDICATED)
      (SERVICE_NAME = ORCLCDB)
    )
  )
```

创建 DBlink，代码如下：

```
SQL> create database link fleet_dblink connect to c##dave identified by oracle using 'dave';
Database link created.
```

设置 LEAD_CDB_URI 参数，代码如下：

```
SQL> show con_name
CON_NAME
------------------------------
CDB$ROOT
SQL> alter database set lead_cdb_uri = 'dblink:fleet_dblink';
Database altered.
SQL> select property_value from database_properties where property_name='LEAD_CDB_URI';
PROPERTY_VALUE
--------------------------------------------------------------
dblink:fleet_dblink
```

告警文件里也会有记录，代码如下：

```
2018-10-25T16:32:28.201940+08:00
alter database set lead_cdb_uri = 'dblink:fleet_dblink'
2018-10-25T16:32:28.308494+08:00
The role of current CDB in the Fleet is: MEMBER
Completed: alter database set lead_cdb_uri = 'dblink:fleet_dblink'
```

在 Lead CDB 的告警文件中，在启用 CDB Fleet 时也会有相应的记录，代码如下：

```
2018-10-25T16:28:24.278374+08:00
alter database set lead_cdb = true
2018-10-25T16:28:24.278774+08:00
The role of current CDB in the Fleet is: LEAD
Completed: alter database set lead_cdb = true
```

3.3.2　查看 CDB Fleet 中的信息

在 Lead CDB 中可以查看 CDB Fleet 中所有 CDB 的信息，比如数据文件信息。

```
#CDB Fleet:
SQL> show pdbs

   CON_ID CON_NAME           OPEN MODE  RESTRICTED
---------- ------------------------------ ---------- ----------
     2 PDB$SEED              READ ONLY  NO
     3 ORCLPDB1              MOUNTED
     6 FLEET             MOUNTED
SQL> alter pluggable database fleet open;
Pluggable database altered.
```

这里必须把 PDB 打开，否则在 Lead CDB 中无法查看该 PDB。

```
#Lead CDB:
SQL> select pdb_name, status from cdb_pdbs;
PDB_NAME STATUS
--------------- --------------------
PDB$SEED NORMAL
ORCLCDB  STUB
DAVE     NORMAL
FLEET        STUB
DAVE_FRESH   NORMAL
DAVE     NORMAL
6 rows selected.
```

在 Lead CDB 中不能对其他 CDB 中的 PDB 进行修改操作，只能进行查询操作，代码如下：

```
SQL> alter session set container=fleet;
ERROR:
ORA-65283: pluggable database FLEET does not belong to the current
```

```
container database
```

3.4 管理 PDB

管理 PDB 和管理普通的 Non-CDB 一样，可以管理表空间、数据文件、临时文件和用户对象等，但是有些任务是无法在 PDB 级别执行的，例如下面的操作。

- 启动和关闭 CDB。
- 修改 CDB。
- 管理以下组件：进程、内存、错误和告警信息、诊断数据、控制文件、在线 redo 日志、归档日志、UNDO 模式。
- 创建、插入、拔出和删除 PDB。

3.4.1 连接 PDB

可以先配置 TNS，然后通过 SQL*PLUS 命令直接访问容器；也可以先连接 CDB，然后通过 ALTER 命令访问 PDB。

（1）通过 TNS 直接访问

TNS 配置代码如下：

```
dave =
  (DESCRIPTION =
    (ADDRESS = (PROTOCOL = TCP)(HOST = 192.168.56.168)(PORT = 1522))
    (CONNECT_DATA =
      (SERVER = DEDICATED)
      (SERVICE_NAME = dave)
    )
  )
```

通过 SQL*PLUS 命令直接访问，代码如下。对于这种访问方式，用户只要有连接该 PDB 的权限就可以了。

```
[oracle@18cDG1 admin]$ sqlplus system/oracle@dave
SQL*Plus: Release 18.0.0.0.0 - Production on Thu Oct 25 05:12:44
2018
Version 18.3.0.0.0
Copyright (c) 1982, 2018, Oracle. All rights reserved.
Last Successful login time: Tue Oct 23 2018 07:41:21 -04:00
Connected to:
Oracle Database 18c Enterprise Edition Release 18.0.0.0.0 -
```

```
Production
    Version 18.3.0.0.0
    SQL> show con_name
    CON_NAME
    ------------------------------
    DAVE
```

如果访问时报如下错误：

```
    ORA-12514: TNS:listener does not currently know of service
requested in connect descriptor
```

可以参考笔者的博客文章来解决：《Oracle 18c 通过 service_name 连接 PDB 报 ORA-12514 错误解决方法》（https://www.cndba.cn/cndba/dave/article/2997）。

（2）通过 ALTER SESSION 命令访问

这种访问方式只能普通用户使用，代码如下：

```
    SQL> show pdbs
       CON_ID CON_NAME          OPEN MODE  RESTRICTED
    ---------- ------------------------------ ---------- ----------
         2 PDB$SEED            READ ONLY  NO
         3 DAVE            READ WRITE NO
    SQL> alter session set container=dave;
    Session altered.
    SQL> show con_name
    CON_NAME
    ------------------------------
    DAVE
```

3.4.2　在系统级别修改 PDB

可以使用 ALTER SYSTEM 命令动态修改 PDB，如果当前容器是 PDB，那么可以执行以下命令：

- ALTER SYSTEM FLUSH { SHARED_POOL | BUFFER_CACHE | FLASH_CACHE }；
- ALTER SYSTEM { ENABLE | DISABLE } RESTRICTED SESSION；
- ALTER SYSTEM SET USE_STORED_OUTLINES；
- ALTER SYSTEM { SUSPEND | RESUME }；
- ALTER SYSTEM CHECKPOINT；
- ALTER SYSTEM CHECK DATAFILES；

- ALTER SYSTEM REGISTER；
- ALTER SYSTEM { KILL | DISCONNECT } SESSION；
- ALTER SYSTEM SET 初始化参数。

对于修改初始化参数，如果 V$SYSTEM_PARAMETER 视图中的列 ISPDB_MODIFIABLE 是 TRUE，那么可以在 PDB 级别进行修改，并不会影响 CDB 的参数值。

在 Oracle 18c 中，共有 210 个可在 PDB 中修改的参数。

```
SQL>select count (*)  from v$system_parameter where
ispdb_modifiable='TRUE';
  COUNT (*)
--------------
    210
```

3.4.3 在数据库级别修改 PDB

在数据库级别修改 PDB，主要是使用 ALTER PLUGGABLE DATABASE 命令，相当于使用 ALTER DATABASE 命令修改 Non-CDB。

如果当前容器是 PDB，则执行 ALTER PLUGGABLE DATABASE 命令的结果是覆盖从 CDB 继承来的默认值，但不会影响其他的容器，相关命令如下。

验证当前容器是 PDB，代码如下：

```
SQL>SHOW CON_NAME
CON_NAME
------------------------------
CNDBAPDB
```

关闭 PDB，代码如下：

```
SQL>ALTER PLUGGABLE DATABASE CLOSE IMMEDIATE;
```

以只读方式打开 PDB，代码如下：

```
SQL>ALTER PLUGGABLE DATABASE OPEN READ ONLY;
```

在线查看数据文件，代码如下：

```
SQL>ALTER PLUGGABLE DATABASE DATAFILE
'/u01/app/oracle/oradata/lei/cndbapdb/lei01.dbf' ONLINE;
```

修改默认表空间，代码如下：

```
SQL>ALTER PLUGGABLE DATABASE DEFAULT TABLESPACE cndba_tbs;
SQL>ALTER PLUGGABLE DATABASE DEFAULT TEMPORARY TABLESPACE
cndba_temp;
```

设置 PDB 的存储大小，代码如下：

```
SQL>ALTER PLUGGABLE DATABASE STORAGE (MAXSIZE 20G);
SQL>ALTER PLUGGABLE DATABASE STORAGE (MAXSIZE UNLIMITED);
SQL>ALTER PLUGGABLE DATABASE STORAGE UNLIMITED;
```

设置强制记录日志，代码如下：

```
SQL>ALTER PLUGGABLE DATABASE NOLOGGING;
SQL>ALTER PLUGGABLE DATABASE ENABLE FORCE LOGGING;
```

3.4.4 启动/关闭 PDB

PDB 默认处于 MOUNT（挂载）模式，和 Non-CDB 的 MOUNT 模式一样，可以进行相关操作。如表 3-3 所示是 PDB 的打开模式及其说明。

表 3-3　PDB 的打开模式及其说明

打开模式	说　明
OPEN READ WRITE	读写模式，允许用户进行读写操作
OPEN READ ONLY	只读模式，只允许用户读取数据，不允许进行修改操作
OPEN MIGRATE	在当前模式，可以执行升级脚本操作（ALTER DATABASE OPEN UPGRADE）
MOUNT	不允许进行任何修改操作，只允许数据库管理员访问，无法读取/修改数据文件。此时内存中关于 PDB 的信息会被移除，可以进行冷备份

在打开 PDB 时，Oracle 会检查 PDB 与 CDB 的兼容性。如果存在不兼容的地方，Oracle 会给出一个警告或错误信息。如果是一个警告，那么警告信息将被记录在告警日志中，PDB 依然可以正常打开，返回结果不会显示告警消息。如果是一个错误，那么在打开 PDB 时会显示一条提示消息，返回结果会提示该错误信息，并将错误信息记录在告警日志中。

当出现错误时，PDB 将以 RESTRICTED 方式打开，只对有 RESTRICTED SESSION 权限的用户开放，其他用户无法连接该 PDB。相关的告警和错误信息可以通过查询 PDB_PLUG_IN_VIOLATIONS 视图来查看。

1．打开 PDB

在 CDB root 中，可以通过 STARTUP PLUGGABLE DATABASE、ALTER PLUGGABLE DATABASE 命令打开一个或多个 PDB，也可以在 PDB 中执行 STARTUP、SHUTDOWN 命令打开和关闭 PDB。这里主要介绍第一种方式。

以读写方式打开 CNDBAPDB，代码如下：

```
SQL>STARTUP PLUGGBLE DATABASE cndbapdb OPEN READ WRITE;
SQL>ALTER PLUGGABLE DATABASE cndbapdb OPEN {READ WRITE};
```

以只读方式打开 CNDBAPDB，代码如下：

```
SQL>STARTUP PLUGGABLE DATABASE cndbapdb OPEN READ ONLY;
SQL>ALTER PLUGGABLE DATABASE cndbapdb OPEN READ ONLY;
```

以升级脚本方式打开 CNDBAPDB，代码如下：

```
SQL>STARTUP PLUAGGBLE DATABASE cndbapdb OPEN UPGRADE;
SQL>ALTER PLUGGABLE DATABASE cndbapdb OPEN UPGRADE;
```

同时打开多个 PDB（cndbapdb、zhixinpdb），代码如下：

```
SQL>ALTER PLUGGABLE DATABASE cndbapdb, zhixinpdb OPEN READ WRITE;
```

打开所有的 PDB，代码如下：

```
SQL>ALTER PLUGGABLE DATABASE ALL OPEN READ WRITE;
```

除 zhixinpdb 外，打开其他的 PDB，代码如下：

```
SQL>ALTER PLUGGABLE DATABASE ALL EXCEPT zhixinpdb OPEN READ WRITE;
```

2．保存当前 PDB 的打开状态

如果想在重启 CDB 时，PDB 能够被自动打开，那么可以在 PDB 被打开时，执行命令保存 PDB 的当前打开状态，这样下次重启 CDB 时，PDB 就会被自动打开。如果不想保存 PDB 的打开状态，则执行 ALTER PLUGGABLE DATABASE cndbapdb DISCARD STATE 命令取消设置即可。

保存一个 PDB 的打开状态，代码如下：

```
SQL>ALTER PLUGGABLE DATABASE cndbapdb SAVE STATE;
```

保存所有 PDB 的打开状态，代码如下：

```
SQL>ALTER PLUGGABLE DATABASE ALL SAVE STATE;
```

保存多个 PDB 的打开状态，代码如下：

```
SQL>ALTER PLUGGABLE DATABASE cndbapdb,zhixinpdb SAVE STATE;
```

除 zhixinpdb 外，保存其他 PDB 的打开状态，代码如下：

```
SQL>ALTER PLUGGABLE DATABASE ALL EXCEPT zhixinpdb SAVE STATE;
```

3. 关闭 PDB

在 CDB root 中可以通过 ALTER PLUGGABLE DATABASE 命令关闭 PDB，代码如下：

```
SQL>ALTER PLUGGABLE DATABASE cndbapdb CLOSE;
```

如果连接到 PDB 中，可以通过 SHUTDOWN 命令关闭 PDB，代码如下：

```
SQL>SHUTDOWN IMMEDIATE;
```

3.5 PDB 快照

Oracle 从 18c 版本开始支持单独对 PDB 进行快照，而不需要对整个 CDB 进行快照。快照是数据库基于时间点的完整镜像，主要用于快速恢复和 PDB 复制。例如，想快速创建一个测试环境，或者 PDB 发生了错误，就可以利用定期产生的快照进行恢复。

利用 PDB 快照创建新 PDB 的语法如下：

SQL>CREATE PLUGGABLE DATABASE cndba FROM cndbapdb USING SNAPSHOT cndbapdb_snap0614;

可以手动生成快照，也可以指定每隔多长时间自动生成一次快照，最多可以存储 8 个快照（由参数 MAX_PDB_SNAPSHOTS 控制）。

注意，截至当前最新的 Oracle 18c 版本，在创建快照时会报以下错误：

```
ORA-12754: Feature PDB SNAPSHOT CAROUSEL is disabled due to missing capability
```

具体解决方法可以参考笔者的博客文章（https://www.cndba.cn/dave/article/2968）。

3.5.1 修改快照个数

查看当前可存储的快照数，代码如下：

```
SQL>SET LINESIZE 150
SQL>COL CON_ID FORMAT 99999
SQL>COL PROPERTY_NAME FORMAT a17
SQL>COL PDB_NAME FORMAT a9
SQL>COL VALUE FORMAT a3
SQL>COL DESCRIPTION FORMAT a43
SQL> SELECT r.CON_ID, p.PDB_NAME, PROPERTY_NAME,
     PROPERTY_VALUE AS value, DESCRIPTION
FROM  CDB_PROPERTIES r, CDB_PDBS p
WHERE  r.CON_ID = p.CON_ID
AND   PROPERTY_NAME LIKE 'MAX_PDB%'
ORDER BY PROPERTY_NAME;
CON_ID PDB_NAME  PROPERTY_NAME      VAL DESCRIPTION
------ --------- --- -------------------------------------------
     4 DAVE    MAX_PDB_SNAPSHOTS 8   maximum number of snapshots for
a given PDB
```

修改快照数需要切换到相应的 PDB 下进行操作，支持动态修改，代码如下：

```
SQL> show pdbs
   CON_ID CON_NAME                 OPEN MODE  RESTRICTED
---------- ------------------------------ ---------- ----------
     2 PDB$SEED             READ ONLY  NO
     4 DAVE             READ WRITE NO
     6 FLEET            MOUNTED
     7 DAVE_FRESH                 MOUNTED
SQL> alter session set container=dave;
Session altered.
SQL> alter pluggable database set max_pdb_snapshots=7;
Pluggable database altered.
```

3.5.2 创建 PDB 快照

可以手动创建快照，也可以设置每过一段时间自动创建快照。

1. 设置自动创建 PDB 快照

可以设置每隔一段时间自动创建一个 PDB 快照，例如每 10 小时、每 30 分钟等，语法如下：

```
EVERY snapshot_interval [MINUTES|HOURS]
```

如果时间是分钟，那么分钟数必须小于 3000；如果时间是小时，那么小时数必须小于 2000。

查看当前 PDB 快照模式，代码如下：

```
SQL>SELECT SNAPSHOT_MODE "S_MODE", SNAPSHOT_INTERVAL/60
"SNAP_INT_HRS" FROM DBA_PDBS;
S_MODE SNAP_INT_HRS
------ ------------
MANUAL
```

设置每隔 24 小时就创建一个快照，需要在相应的 PDB 下执行命令，代码如下：

```
SQL>ALTER PLUGGABLE DATABASE SNAPSHOT MODE EVERY 24 HOURS;
```

再次查看快照模式，代码如下：

```
SQL>SELECT SNAPSHOT_MODE "S_MODE", SNAPSHOT_INTERVAL/60
"SNAP_INT_HRS" FROM DBA_PDBS;
S_MODE  SNAP_INT_HRS
-------------- ------------------
AUTO      24
```

2. 手动创建 PDB 快照

在创建快照时，如果不指定快照名称，那么 Oracle 会自动生成一个名称，代码如下：

```
SQL> show con_name
CON_NAME
------------------------------
DAVE
SQL> alter pluggable database snapshot dave_20181025205800;
Pluggable database altered.
```

查看快照信息，代码如下：

```
SQL>SET LINESIZE 150
SQL>COL CON_NAME FORMAT a10
SQL>COL SNAPSHOT_NAME FORMAT a30
SQL>COL SNAP_SCN FORMAT 9999999
SQL>COL FULL_SNAPSHOT_PATH FORMAT a55
SQL>SELECT CON_ID, CON_NAME, SNAPSHOT_NAME,
     SNAPSHOT_SCN AS snap_scn, FULL_SNAPSHOT_PATH
FROM  DBA_PDB_SNAPSHOTS
ORDER BY SNAP_SCN;
CON_ID CON_NAME   SNAPSHOT_NAME         SNAP_SCN
FULL_SNAPSHOT_PATH
-------------------------------------------------------------
```

```
        4 DAVE     SNAP_2631944658_990478614     2121867
/opt/oracle/oradata/ORCLCDB/snap_2631944658_2121867.pdb
        4 DAVE     DAVE_20181025205800      2122260
/opt/oracle/oradata/ORCLCDB/snap_2631944658_2122260.pdb
```

3.5.3 删除快照

可以一个一个地删除快照，也可以一次性删除所有快照。

删除一个快照的代码如下：

```
SQL>ALTER PLUGGABLE DATABASE DROP SNAPSHOT DAVE_20181025205800;
```

在删除所有快照时，将 MAX_PDB_SNAPSHOTS 参数设置为 0 就会自动删除该 PDB 的所有快照，删除速度也快很多，代码如下：

```
SQL>ALTER PLUGGABLE DATABASE SET MAX_PDB_SNAPSHOTS=0;
```

3.6 监控 CDB 和 PDB

Oracle 18c 和之前的版本一样，有大量的视图可以用来查看 CDB 和 PDB 的信息。具体视图如表 3-4 所示。

表 3-4 Oracle 18c 中的视图

视　　图	说　　明
容器数据对象，包括： • V$视图； • GV$视图； • CDB_视图； • DBA_HIST*视图	容器数据对象用于显示关于多租户的相关信息。每个容器数据对象通过 CON_ID 来区分不同的容器
{CDB|DBA}_PDBS	显示与 CDB 相关的 PDB 的信息，包括每个 PDB 的状态
CDB_PROPERTIES	显示 CDB 中每个容器的属性
{CDB|DBA}_PDB_HISTORY	显示每个 PDB 的历史记录信息
{CDB|DBA}_HIST_PDB_INSTANCE	显示 PDB 和实例的负载信息库
{CDB|DBA}_PDB_SAVED_STATES	显示 CDB 中当前保存的 PDB 的状态信息
{CDB|DBA}_APPLICATIONS	记录应用程序容器中的所有应用程序
{CDB|DBA}_APP_STATEMENTS	记录应用程序容器中应用程序安装、升级和补丁操作的所有语句

续表

视　图	说　明
{CDB|DBA}_APP_PATCHES	记录应用程序容器中所有应用程序的补丁信息
{CDB|DBA}_APP_ERRORS	记录应用程序容器中所有应用程序的错误信息
{CDB|DBA}_CDB_RSRC_PLANS	显示所有关于 CDB 资源计划的信息
{CDB|DBA}_CDB_RSRC_PLAN_DIRECTIVES	显示所有关于 CDB 资源计划指令的信息
PDB_ALERTS	记录 PDB 告警原因的描述
PDB_PLUG_IN_VIOLATIONS	显示 PDB 与其所属的 CDB 之间不兼容的信息。此视图还显示执行 DBMS_PDB.CHECK_PLUG_COMPATIBILITY 命令生成的信息
{USER|ALL|DBA|CDB}_OBJECTS	显示数据库的对象信息
{ALL|DBA|CDB}_SERVICES	显示数据库 Service 的信息，并且 PDB 列显示的是与 PDB 相关的 Service
[G]V$CONTAINERS	显示与当前 CDB 相关的信息，包括 root 和所有 PDB
[G]V$PDBS	显示与当前 CDB 相关的所有 PDB 的信息，包括 PDB 的打开模式
[G]V$PDB_INCARNATION	显示所有 PDB Incarnation 的信息。每当使用 RESETLOGS 选项打开 PDB 时，Oracle 都会创建一个新的 PDB Incarnation
[G]V$SYSTEM_PARAMETER [G]V$PARAMETER	显示关于初始化参数的信息，其中列 ISPDB_MODIFIABLE 的值表示该参数是否可以在 PDB 中修改
V$DIAG_ALERT_EXT [G]V$DIAG_APP_TRACE_FILE [G]V$DIAG_OPT_TRACE_RECORDS V$DIAG_SESS_OPT_TRACE_RECORDS V$DIAG_SESS_SQL_TRACE_RECORDS [G]V$DIAG_SQL_TRACE_RECORDS [G]V$DIAG_TRACE_FILE [G]V$DIAG_TRACE_FILE_CONTENTS	显示 CDB 中当前容器的 Trace 文件和告警日志信息
V$DIAG_INCIDENT V$DIAG_PROBLEM	显示 CDB 中关于当前容器的问题和故障的信息

3.6.1　查看 CDB 中容器的信息

可以通过 V$CONTAINERS 视图来查看容器的信息，代码如下：

```
SQL>COLUMN NAME FORMAT A8
SQL>SELECT NAME, CON_ID, DBID, CON_UID, GUID FROM V$CONTAINERS
ORDER BY CON_ID;
NAME     CON_ID     DBID     CON_UID GUID
-------- ---------- ---------- --------------------------------
CDB$ROOT  1  2161911720     6E5AE8B7136378DBE053B201A8C02AC3
PDB$SEED       2  247385757      247385757
6E5AE8B7136478DBE053B201A8C02AC3
CNDBAPDB  4  1757102921      948422718
6F0AA35EFEA45427E053B201A8C0A225
```

3.6.2 查看 PDB 的信息

可以通过 CDB_PDBS 视图来查看关于 PDB 的相关信息，代码如下：

```
SQL>COLUMN PDB_NAME FORMAT A15
SQL>SELECT PDB_ID, PDB_NAME, STATUS FROM DBA_PDBS ORDER BY PDB_ID;
   PDB_ID PDB_NAME      STATUS
---------- --------------- --------------------
     2 PDB$SEED    NORMAL
     4 CNDBAPDB    NORMAL
```

3.6.3 查看 PDB 的打开状态和打开时间

可以通过 V$PDBS 视图来查看 PDB 的打开状态和打开时间，代码如下：

```
SQL>COLUMN NAME FORMAT A15
SQL>COLUMN RESTRICTED FORMAT A10
SQL>COLUMN OPEN_TIME FORMAT A40
SQL>SELECT NAME, OPEN_MODE, RESTRICTED, OPEN_TIME FROM V$PDBS;
NAME     OPEN_MODE        RESTRICTED OPEN_TIME
--------------- -------------------- ----------
----------------------------------------
PDB$SEEDREAD ONLY         NO   23-JUN-18 10.36.21.033 AM +08:00
CNDBAPDBREAD WRITE        NO   23-JUN-18 10.31.02.091 PM +08:00
```

3.6.4 查看 PDB 中的表

查看所有 PDB 中用户 CNDBA 和用户 LEI 的表，其中，设置 PDB_ID>2 是为了不查询 CDB root 和 PDB seed，代码如下：

```
SQL>COLUMN PDB_NAME FORMAT A15
```

```
SQL>COLUMN OWNER FORMAT A15
SQL>COLUMN TABLE_NAME FORMAT A30
SQL>SELECT p.PDB_ID, p.PDB_NAME, t.OWNER, t.TABLE_NAME
  FROM DBA_PDBS p, CDB_TABLES t
  WHERE p.PDB_ID > 2 AND
       t.OWNER IN ('LEI','CNDBA') AND
       p.PDB_ID = t.CON_ID
  ORDER BY p.PDB_ID;
    PDB_ID PDB_NAME   OWNER    TABLE_NAME
---------- -------------- ---------------
------------------------------
     3 HRPDB        LEI         EMPLOYEES
     3 HRPDB        LEI         JOBS
     4 CNDBAPDB     CNDBA       PRODUCT_REF_LIST_NESTEDTAB
     4 CNDBAPDB     CNDBA       PROMOTIONS
     4 CNDBAPDB     CNDBA       PRODUCT_DESCRIPTIONS
```

3.6.5 查看 PDB 的数据文件

查看所有与 PDB 相关的数据文件，包括 PDB seed，代码如下：

```
SQL>COLUMN PDB_ID FORMAT 999
SQL>COLUMN PDB_NAME FORMAT A8
SQL>COLUMN FILE_ID FORMAT 9999
SQL>COLUMN TABLESPACE_NAME FORMAT A10
SQL>COLUMN FILE_NAME FORMAT A40
SQL>SELECT p.PDB_ID, p.PDB_NAME, d.FILE_ID, d.TABLESPACE_NAME,
d.FILE_NAME
  FROM DBA_PDBS p, CDB_DATA_FILES d
  WHERE p.PDB_ID = d.CON_ID
  ORDER BY p.PDB_ID;
PDB_ID PDB_NAME FILE_ID TABLESPACE FILE_NAME
------------------------------------------------------
2 PDB$SEED 6 SYSAUX /u01/app/opracle/oradata/pdbseed/SYSAUX.dbf
2 PDB$SEED 5 SYSTEM /u01/app/opracle/oradata/pdbseed/SYSTEM.dbf
3 PDB2     9 SYSAUX /u01/app/opracle/oradata/pdb2/SYSAUX.dbf
3 PDB2     8 SYSTEM /u01/app/opracle/oradata/pdb2/SYSTEM.dbf
3 PDB2    13 USER   /u01/app/opracle/oradata/pdb2/hrpdb_usr.dbf
4 CNDBAPDB  15 SYSTEM /u01/app/opracle/oradata/cndbapdb/SYSTEM.dbf
4 CNDBAPDB 16 SYSAUX  /u01/app/opracle/oradata/cndbapdb/SYSAUX.dbf
4 CNDBAPDB 18 USER /u01/app/opracle/oradata/cndbapdb/USER01.dbf
```

3.6.6 使用 CONTAINERS 命令跨 PDB 查询

可以使用 CONTAINERS 命令查询 CDB 中所有容器（会忽略以 RESTRICT 模式打开的 PDB）的表和视图。其中，初始化参数 CONTAINERS_PARALLEL_DEGREE 用来控制查询的并行数，默认是 65 535，如果指定的值小于 65 535，就使用指定的值。

计算公式如下：

```
CONTAINERS_PARALLEL_DEGREE=max（min（cpu_count,number_of_open_containers）,#instances）
```

如果在查询中某个 PDB 报错，那么查询结果中就不会包含该 PDB，而且不会提示错误信息。

查询所有 PDB 中的表，代码如下：

```
SQL>SELECT * FROM CONTAINERS（CNDBA）;
```

查询指定 PDB 中的表，代码如下：

```
SQL>SELECT * FROM CONTAINERS（CNDBA） WHERE CON_ID IN（3,4）;
```

3.7 CDB 用户、PDB 用户及角色管理

多租户架构中的用户管理和 Non-CDB 数据库中的用户管理有所区别，新增了两个概念，即公共用户（Common User）和本地用户（Local User），与之对应的角色是公共角色（Common Role）和本地角色（Local Role）。

3.7.1 用户

公共用户可以在 CDB 和 PDB 中同时存在，能够连接 root 和 PDB 并进行操作；而本地用户只在特定的 PDB 中存在，也只能在特定的 PDB 中进行操作。在 PDB 中不能创建公共用户，在 CDB 中（CDB$ROOT 中）不能创建本地用户。

1. 公共用户

在 CDB 中创建公共用户需要以 C## 开头（由参数 COMMON_USER_PREFIX 控制），而在 Non-CDB 中用户名不能以 C##开头，

用户名不区分大小写。

错误语法：

```
SQL>CREATE USER LEI IDENTIFIED BY ORACLE CONTAINER=ALL;
CREATE USER LEI IDENTIFIED BY ORACLE
            *
ERROR AT LINE 1:
ORA-65096: INVALID COMMON USER OR ROLE NAME
```

正确语法：

```
SQL>CREATE USER C##LEI IDENTIFIED BY ORACLE CONTAINER=ALL;
USER CREATED.
```

当公共用户创建成功后，默认情况下会在 CDB 中的所有 PDB 中创建公共用户，代码如下：

```
    SQL>col username for a20
    SQL>select username,common,con_id from cdb_users where username
like '%LEI%';
    USERNAME     COM     CON_ID
    --------------------------- --------------- ----------
    C##LEI            YES          1
    C##LEI            YES          3
```

注意，以 C##开头是由参数 COMMON_USER_PREFIX 控制的，如果该参数值为空，则没有该要求，但是 Oracle 不建议这么做。

```
    SQL>show parameter COMMON_USER_PREFIX
    NAME                    TYPE  VALUE
    ------------------------------------ -----------
------------------------------
    common_user_prefix            string    C##
```

2. 本地用户

本地用户和 Non-CDB 中的用户没有区别。只能在 PDB 中创建本地用户，无法在 CDB root 中创建本地用户。本地用户无法登录其他 PDB 或 root，在不同的 PDB 中本地用户的用户名可以相同。和公共用户名的名称必须以 C##开头不同，创建本地用户时名称不能以 C##开头。

错误语法：

```
SQL>create user c##expect identified by oracle container=CURRENT;
```

```
create user c##expect identified by oracle container=CURRENT
            *
ERROR at line 1:
ORA-65094: invalid local user or role name
```

正确语法：

```
SQL>create user expect identified by oracle container=CURRENT;
User created.
```

3.7.2 角色

角色是一些权限的集合。一个角色里包含了各种权限，主要是为了管理方便，不用一个一个地赋予或撤回。从 Oracle 12c 开始，用户创建的角色要么是公共角色，要么是本地角色。和用户一样，公共角色对所有 PDB 都有效，而本地角色只对指定的 PDB（创建角色所在的 PDB）有效，并且公共角色的名称也必须以 C##开头，受参数 COMMON_USER_PREFIX 的控制。

只要用户和角色在容器中同时存在，就可以正常地将角色赋给相应的用户，而不管是公共用户还是本地用户。其他 PDB 中的本地角色不能在另一个 PDB 中使用，因为该角色只存在于指定的 PDB 中。

1. 公共角色

公共角色存在于 CDB root 或应用程序 root 中，对其中所有的 PDB 都有效。公共角色分为 Oracle 自带和用户自定义两种。

（1）Oracle 自带：例如 DBA、PUBLIC 等角色。

（2）用户自定义：用户通过 CREATE ROLE … CONTAINER=ALL 语句在 CDB root 或应用程序 root 中创建的角色。角色的名称必须唯一且以 C##开头。相关操作如下。

确认当前容器是 root，代码如下：

```
SQL> show con_name
CON_NAME
------------------------------
CDB$ROOT
```

名称不以 C##开头来创建角色，会报如下错误：

```
SQL> CREATE ROLE suyi_role;
```

```
CREATE ROLE suyi_role
        *
ERROR at line 1:
ORA-65096: invalid common user or role name
```

正常创建角色并给角色赋权限，代码如下：

```
SQL> CREATE ROLE C##SUYI_LEI;
Role created.
SQL> grant dba,resource,connect to c##suyi_lei;
Grant succeeded.
```

将公共角色赋给公共用户，代码如下：

```
SQL> grant c##suyi_lei to c##lei;
Grant succeeded.
```

查看角色是否是公共角色，代码如下：

```
SQL> select role,common from dba_roles where role='C##SUYI_LEI';
ROLE                 COMMON
-----------------------------------------------------------------
C##SUYI_LEI            YES
```

创建一个和已存在的角色名称相同的角色，代码如下：

```
SQL> CREATE ROLE C##SUYI_LEI;
CREATE ROLE C##SUYI_LEI
        *
ERROR at line 1:
ORA-01921: role name 'C##SUYI_LEI' conflicts with another user or
role name
```

2. 本地角色

本地角色仅存在于单个 PDB 中，即创建该角色的 PDB 中，其他 PDB 无法使用该角色。本地角色的名称在不同的 PDB 中可以是一样的，也就是说，两个 PDB 中的本地角色的名称可以是相同的。实际上从数据库的角度也很好理解，多租户的概念主要就是为了区分 PDB，各个 PDB 之间是完全独立的。

创建本地角色并赋权限，其语法如下：

```
SQL> alter session set container=cndbapdb;
Session altered.
SQL> create role suyi;
Role created.
```

```
SQL> grant dba,resource,connect to suyi;
Grant succeeded.
```

查看角色属性，代码如下：

```
SQL> col role for a20
SQL> select role,common from dba_roles where role='SUYI';
ROLE     COM
-------------------- -----------
SUYI         NO
```

3.8 管理 CDB 和 PDB 的表空间

Oracle 多租户架构下的表空间管理和之前讲解的 Non-CDB 数据库中的表空间管理没有太大区别。对于表空间的数据文件的存放位置，Oracle 没有要求，只要有权限即可。但是为了规范，PDB 的数据文件应该和 CDB root 的数据文件分开存放，以便管理。

与 Non-CDB 相比，CDB 中的表空间主要有以下特点。

- 一个永久表空间只能与一个容器相关联。
- 在当前容器中创建表空间时，表空间将与该容器相关联。
- 当 CDB 禁用本地 UNDO 模式时，CDB 只能有一个活动的 UNDO 表空间，或者 Oracle RAC CDB 的每个节点各有一个活动的 UNDO 表空间。当为 CDB 启用本地 UNDO 模式时，CDB 中的每个容器都有自己的 UNDO 表空间。
- 在 Oracle RAC 集群中，每个节点都需要一个本地 UNDO 表空间。
- CDB 中的每个容器都有自己的默认临时表空间，包括 CDB root、每个 PDB、每个应用程序 root 和每个应用程序 PDB。

CDB 中的每个容器都有自己的默认临时表空间或表空间组。可以为单个容器创建临时表空间，并可以将容器中的特定用户分配给这些临时表空间。当拔掉 PDB 时，它的临时表空间也会被删除。当用户没有在容器中显式地分配临时表空间时，用户的临时表空间是容器的默认临时表空间。

3.8.1　管理 CDB 表空间

Oracle 18c 中的 CDB 表空间管理和 Oracle 之前版本的表空间管理没有任何区别。先连接到 CDB root，然后进行相关操作。

查看默认表空间，代码如下：

```
SQL>SET LINESIZE 200
SQL>COL PROPERTY_NAME FORMAT A30
SQL>COL PROPERTY_VALUE FORMAT A20
SQL>SELECT PROPERTY_NAME,PROPERTY_VALUE FROM DATABASE_PROPERTIES
WHERE PROPERTY_NAME IN
('DEFAULT_PERMANENT_TABLESPACE','DEFAULT_TEMP_TABLESPACE');
PROPERTY_NAME            PROPERTY_VALUE
------------------------------ --------------------
DEFAULT_PERMANENT_TABLESPACE   USERS
DEFAULT_TEMP_TABLESPACE       TEMPTS1
```

创建临时表空间，代码如下：

```
SQL>create temporary tablespace temp01 tempfile
'/u01/app/oracle/oradata/lei/temp02.dbf' size 5m autoextend off;
```

创建永久表空间，代码如下：

```
SQL>CREATE TABLESPACE cdb_users DATAFILE '/ u01
/app/oracle/oradata/lei/cdb_users01.dbf' SIZE 5M autoextend off
segment space management auto extent management local;
```

创建 UNDO 表空间，代码如下：

```
SQL>create undo tablespace UNDOTBS2 datafile
'/u01/app/oracle/oradata/lei/undotbs02' size 5m reuse autoextend off
extent management local;
```

指定默认表空间，代码如下：

```
SQL>ALTER DATABASE DEFAULT TABLESPACE cdb_users;
```

指定默认临时表空间，代码如下：

```
SQL>ALTER DATABASE DEFAULT TEMPORARY TABLESPACE temp01;
```

添加数据文件、修改默认表空间等操作的语法和 Oracle 之前版本的相关语法没有区别，这里不再一一列举。

3.8.2 管理 PDB 表空间

管理 PDB 表空间的大部分操作和管理 CDB 表空间的操作类似，只有指定默认表空间的语法不同。

连接到 PDB，代码如下：

```
SQL>ALTER SESSION SET CONTAINER=CNDBAPDB;
```

或

```
SQL>conn c##pdbadmin/oracle@cndbapdb;
```

创建临时表空间，代码如下：

```
SQL>CREATE TEMPORARY TABLESPACE P_TEMP01 TEMPFILE
'/u01/app/oracle/oradata/lei/temp02.dbf' SIZE 5M AUTOEXTEND OFF;
```

创建永久表空间，代码如下：

```
SQL>CREATE TABLESPACE pdb_users DATAFILE
'/u01/app/oracle/oradata/lei/cdb_users01.dbf' SIZE 5M AUTOEXTEND OFF
SEGMENT SPACE MANAGEMENT AUTO EXTENT MANAGEMENT LOCAL;
```

创建 UNDO 表空间，代码如下：

```
SQL>CREATE UNDO TABLESPACE P_UNDOTBS2 DATAFILE
'/u01/app/oracle/oradata/lei/undotbs02' SIZE 5M REUSE AUTOEXTEND OFF
EXTENT MANAGEMENT LOCAL;
```

查看 PDB 的表空间，代码如下：

```
SQL>SELECT p.PDB_ID, p.PDB_NAME, d.TABLESPACE_NAME FROM DBA_PDBS
p, CDB_TABLESPACES d  WHERE p.PDB_ID = d.CON_ID ORDER BY p.PDB_ID;
    PDB_ID PDB_NAME   TABLESPACE_NAME
---------- ---------- ------------------------------
         3 CNDBAPDB   SYSTEM
         3 CNDBAPDB   SYSAUX
         3 CNDBAPDB   UNDOTBS1
         3 CNDBAPDB   TEMP
         3 CNDBAPDB   USERS
         3 CNDBAPDB   PDB_USERS
         3 CNDBAPDB   P_TEMP01;
```

指定默认表空间，代码如下：

```
SQL>ALTER PLUGGABLE DATABASE DEFAULT TABLESPACE pdb_users;
```

或

```
SQL>ALTER DATABASE DEFAULT TABLESPACE pdb_users;
```

指定默认临时表空间，代码如下：

```
SQL>ALTER PLUGGABLE DATABASE DEFAULT TEMPORARY TABLESPACE
p_temp01;
```

或

```
SQL>ALTER DATABASE DEFAULT TEMPORARY TABLESPACE p_temp01;
```

查看当前 PDB 的默认永久表空间和临时表空间，代码如下：

```
SQL>SET LINESIZE 200
SQL>COL PROPERTY_NAME FORMAT A30
SQL>COL PROPERTY_VALUE FORMAT A20
SQL>SELECT PROPERTY_NAME,PROPERTY_VALUE FROM DATABASE_PROPERTIES
WHERE PROPERTY_NAME IN
('DEFAULT_PERMANENT_TABLESPACE','DEFAULT_TEMP_TABLESPACE');
PROPERTY_NAME              PROPERTY_VALUE
------------------------------ --------------------
DEFAULT_PERMANENT_TABLESPACE   PDB_USERS
DEFAULT_TEMP_TABLESPACE        P_TEMP01
```

3.8.3 查看表空间的使用情况

查看表空间的使用情况，不管是 CDB，还是 PDB，都是通过同一个视图来查看的，在查询时切换到相应的容器下执行命令即可，代码如下（以下脚本中包含了临时表空间）：

```
SQL>SELECT d.tablespace_name "Name", d.status "Status",
      TO_CHAR (NVL (a.BYTES / 1024 / 1024, 0) , '99,999,990.90')
"Size (M) ",
        TO_CHAR (NVL (a.BYTES - NVL (f.BYTES, 0) , 0) / 1024 /
1024,
              '99999999.99'
             ) USE,
      TO_CHAR (NVL ((a.BYTES - NVL (f.BYTES, 0)) / a.BYTES * 100,
0),
            '990.00'
           ) "Used %"
  FROM SYS.dba_tablespaces d,
     (SELECT tablespace_name, SUM (BYTES) BYTES
```

```
            FROM dba_data_files
        GROUP BY tablespace_name) a,
        (SELECT  tablespace_name, SUM (BYTES) BYTES
            FROM dba_free_space
        GROUP BY tablespace_name) f
 WHERE d.tablespace_name = a.tablespace_name (+)
   AND d.tablespace_name = f.tablespace_name (+)
   AND NOT (d.extent_management LIKE 'LOCAL' AND d.CONTENTS LIKE
'TEMPORARY')
UNION ALL
SELECT d.tablespace_name "Name", d.status "Status",
       TO_CHAR (NVL (a.BYTES / 1024 / 1024, 0), '99,999,990.90')
"Size (M)",
          TO_CHAR (NVL (t.BYTES, 0) / 1024 / 1024, '99999999.99')
USE,
       TO_CHAR (NVL (t.BYTES / a.BYTES * 100, 0), '990.00') "Used %"
  FROM SYS.dba_tablespaces d,
        (SELECT  tablespace_name, SUM (BYTES) BYTES
            FROM dba_temp_files
        GROUP BY tablespace_name) a,
        (SELECT  tablespace_name, SUM (bytes_cached) BYTES
            FROM v$temp_extent_pool
        GROUP BY tablespace_name) t
 WHERE d.tablespace_name = a.tablespace_name (+)
   AND d.tablespace_name = t.tablespace_name (+)
   AND d.extent_management LIKE 'LOCAL'
   AND d.CONTENTS LIKE 'TEMPORARY';
```

结果：

```
Name      Status     Size (M)        USE         Used %
------------------- -------- -----------------------------------
SYSTEM    ONLINE     700.00         438.63     62.66
SYSAUX    ONLINE     550.00         451.63     82.11
UNDOTBS1  ONLINE     200.00         160.44     80.22
USERS     ONLINE     500.00           1.00      0.20
TEMPTS1   ONLINE      20.00           6.00     30.00
```

3.9 PDB 资源管理

Instance Caging（实例化限制）是 Oracle 11g 引入的技术，可以保证同台主机上多个实例对 CPU 资源的合理分配，预防过度负载造成的操作系统不稳定，以及后台进程无法得到 CPU 资源而导致的后台进程异常，比如 LGWR（性能低下）、PMON（数据库不稳定）、LMS（集群脑裂）等。

Oracle 从 12c 开始提出了云的概念。在 CDB 架构中，可以创建多个 PDB，从某种角度来看，每个 PDB 都是一个独立的实例。当一个 CDB 中有多个 PDB 时，可以设置实例化限制来保证这些 PDB 合理地使用 CPU 资源。实例化限制和 Oracle 资源管理器一起被使用可以支持跨实例管理资源分配。

实例化限制使用初始化参数来限制实例可以同时使用的 CPU 数量，也可以限制每个实例最多可以使用的 CPU 数量，从而避免某个实例使用过多的 CPU 资源而影响其他数据库实例的性能。而在此之前，用户是无法控制某个数据库实例可以使用多少 CPU 资源的，完全由数据库实例的负载和操作系统决定。

例如，服务器有 16 个 CPU，上面运行着 4 个数据库实例，那么可以使用实例化限制将 4 个实例可使用的最大 CPU 数量限制为 4 个，这样各个实例之间的影响就会小很多。

通常有以下两种使用实例化限制的方式。

- 平均分配：所有实例可以使用的最多 CPU 数量加起来等于服务器的总 CPU 个数，所以各个实例之间几乎没有什么影响。这种方式适用于对生产环境要求很高和需要稳定性能的情况。
- 过度分配：过度分配就是每个实例可以使用的最大 CPU 数量加起来大于服务器总的 CPU 数量。这种方式适用于测试环境或负载很低的非核心生产环境。

实例化限制需要配合资源计划来使用，在资源计划中指定 CPU 利用率，通过百分比来计算实例可使用的 CPU 资源。例如，在有 4 个 CPU 的服务器上，在资源计划中设置了该消费组的 CPU 使用率是 50%，那么该消费组可以使用的最大 CPU 数量是 2。

另外，这里要注意一点，现在 CPU 都有超线程，但是第二线程的性能和第一线程的性能有差距，所以在估算资源的时候，要适当留一点冗余量，具体说明可参考博客文章《CPU 超线程使用率说明》（https://www.cndba.cn/Expect-le/article/2900）。

3.9.1 启用实例化限制

启用实例化限制非常简单，创建一个具有 CPU 指令的资源计划并设置 CPU_COUNT 初始化参数即可。

（1）使用资源管理器创建一个具有 CPU 指令的资源计划。

（2）设置 CPU_COUNT 参数。

CPU_COUNT 的默认值为 0。一定要根据实际的 CPU 数量进行设置，因为 Oracle 首先会根据该参数值进行 CPU 资源分配，而不管真实的 CPU 数量是多少。当该参数值为 0 时，Oracle 会去查询真实的 CPU 数量。该参数支持在 PDB 级别修改，也支持动态修改，代码如下：

```
SQL> ALTER SYSTEM SET CPU_COUNT = 4;
SQL> SHOW PARAMETER CPU_COUNT
NAME                  TYPE         VALUE
---------------------------- ---- ----------------------------
cpu_count             integer         4
```

3.9.2 多租户环境下的资源管理器

正常情况下，在 CDB 中，多个 PDB 的工作负载会与操作系统、CDB 资源进行争用，而资源计划可以完美地解决该问题，下面就介绍使用资源计划来限制 PDB 的资源使用。

在多租户环境下，资源管理器支持在两个级别进行管理。

- CDB 级别：可以指定如何分配资源给 PDB，以及指定 PDB 的资源分配。
- PDB 级别：可以单独管理某个 PDB 中的负载。

默认情况下，资源管理器在 CDB root 中是启用的。

在多租户环境下使用资源管理器，必须满足以下条件：

- CDB 一定存在且其中要有 PDB；
- 具有 ADMINISTER_RESOURCE_MANAGER 权限。

多租户环境下的资源管理器可以实现以下功能：

- 指定不同的 PDB 分配不同的资源；
- 限制指定 PDB 的 CPU 使用；

- 限制指定 PDB 可以使用的并行执行服务器(Parallel Execution Servers)的数量；
- 限制指定 PDB 的内存使用；
- 指定 PDB 分配最低可用内存；
- 指定 PDB 可以使用的最大内存；
- 限制连接到 PDB 的不同会话的资源使用大小；
- 限制指定 PDB 产生的 I/O；
- 监视 PDB 的资源使用情况。

1．资源计划指令

CDB 资源计划根据其中的资源计划指令将资源分配给 PDB，所以 CDB 资源计划和资源计划指令是父子关系。可以同时为同一个 CDB 中的 PDB 和 PDB 性能概要（PDB Performance Profiles）指定计划指令，但是对于一个当前正在使用的资源计划来说，不能同时有两个计划指令来限制同一个 PDB 和 PDB 性能概要。

（1）PDB 性能概要

PDB 性能概要是给一组具有相同优先级和使用资源的 PDB 设置的资源计划指令。例如，有 3 个 PDB 性能概要分别是“重要”“一般”“无所谓”，PDB 可以通过 DB_PERFORMANCE_PROFILE 初始化参数来指定使用哪个概要。

（2）资源计划指令

资源计划指令用于控制 CPU 和并行执行服务器资源的分配。资源计划指令可以根据概要中指定的每个 PDB 的 share 值来分配相应的资源，share 值越大，分配的资源越多。也可以为 PDB 和性能概要指定 utilization_ limits(使用限制)，该值是一个百分比，用于指定可使用的 CPU 资源。

2．多租户资源管理相关的初始化参数

在多租户环境中，可以通过设置初始化参数来控制 PDB 使用的资源，例如内存、I/O。

（1）内存相关的参数

在多租户环境中，由于所有 PDB 和 CDB root 共享内存，所以内存相关参数的设置要根据实际情况进行分配，并且要根据运行情况及时调整，否则可能会出现一个 PDB 把内存资源用光，导致其他 PDB 无法正常运行的情况。

在设置相关参数之前，多租户环境下必须满足以下条件：

- NONCDB_COMPATIBLE 参数必须设置为 false；
- MEMORY_TARGET 参数设置为 0。

如表 3-5 所示是多租户环境下参数的设置建议。

表 3-5　多租户环境下参数的设置建议

初始化参数	说　明
DB_CACHE_SIZE	PDB 中设置的高速缓冲区大小。 如果设置了 DB_CACHE_SIZE，那么必须满足以下条件： • 参数值必须小于或等于 CDB root 中该参数值的 50%； • 所有 PDB 中该参数值的总和必须小于或等于 CDB root 中该参数值的 50%。 如果设置了参数 SGA_TARGET，那么就需要满足以下条件： • 参数 DB_CACHE_SIZE 和 SHARD_POOL_SIZE 的大小加起来必须小于或等于该 PDB 中参数 SGA_TARGET 大小的 50%； • 单个 PDB 中的参数 DB_CACHE_SIZE 和 SHARD_POOL_SIZE 的大小加起来必须小于或等于 CDB root 中参数 SGA_TARGET 大小的 50%； • 所有 PDB 中的参数 DB_CACHE_SIZE 和 SHARD_POOL_SIZE 的大小加起来必须小于或等于 CDB root 中参数 SGA_TARGET 大小的 50%
SHARD_POOL_SIZE	PDB 中设置的共享池大小。和 DB_CACHE_SIZE 一样，单个 PDB 的参数值或所有 PDB 参数值总和都不能大于 CDB root 中该参数值的 50%
SGA_MIN_SIZE	PDB 中设置的最小的 SGA 大小。如果要设置该参数，则必须满足以下条件： • 其参数值不能大于 CDB root 中 SGA_TARGET 参数值的 50%； • 其参数值不能大于同一个 PDB 中 SGA_TARGET 参数值的 50%； • 所有 PDB 中该参数值总和不能大于 CDB root 中 SGA_TARGET 参数值的 50%。 如果 CDB root 中没有将参数 SGA_TARGET 或参数 SGA_TARGET 的值设置为 0，则没有以上限制
SGA_TARGET	PDB 中设置的最大的 SGA 大小。只有当 SGA_TARGET 初始化参数在 CDB root 中被设置为非 0 时，Oracle 才会强制设置 PDB 中的 SGA_TARGET 参数，并且 PDB 中的 SGA_TARGET 参数值必须不大于 CDB root 中的 SGA_TARGET 的参数值
PGA_AGGREGATE_LIMIT	PDB 中设置的最大的 PGA 大小。 如果要设置该参数，则必须满足以下条件： • 其参数值必须不大于 CDB root 中 PGA_AGGREGATE_LIMIT 的参数值； • 其参数值必须大于或等于同一个 PDB 中 PGA_AGGREGATE_ TARGET 参数值的两倍

续表

初始化参数	说　　明
PGA_AGGREGATE_TARGET	PDB 中设置的总 PGA 的大小。如果要设置该参数，则必须满足以下条件： • 其参数值不能大于 CDB root 中该参数值的大小； • 其参数值不能大于 CDB root 中 PGA_AGGREGATE_LIMIT 参数值的 50%； • 其参数值不能大于同一个 PDB 中 PGA_AGGREGATE_LIMIT 参数值的 50%

（2）I/O 相关的参数

限制 PDB 产生磁盘 I/O 的两个参数分别是 MAX_IOPS 和 MAX_MBPS。大量的磁盘 I/O 会导致性能下降。如果一个 PDB 产生大量的磁盘 I/O，那么就会降低同一个 CDB 中其他 PDB 的性能。

MAX_IOPS：限制每秒产生的 I/O 操作数。

MAX_MBPS：限制每秒产生的 I/O 数据量大小（MB）。

注意，这两个参数对于 Exdata 不生效。

默认情况下，如果在 CDB root 中设置该参数值，那么该 CDB 中的所有 PDB 都继承该参数值。同样，应用程序 root 也是如此。如果在 PDB 中设置了新的参数值，那么原来 CDB root 中的参数值对该 PDB 不再生效，将启用新的参数值。

这两个参数的默认值均为 0，也就是对 PDB 的 I/O 操作是没有限制的。但注意，即使启用了 I/O 限制，一些重要文件的操作仍不在此限制中，例如对控制文件、密码文件的 I/O 操作，这些操作被称为重要 I/O 操作。

可以使用 DBA_HIST_RSRC_PDB_METRIC 视图为 PDB 计算合理的 I/O 限制值。在计算极限值时，需要考虑以下列中的值：IOPS、IOMBPS、IOPS_THROTTLE_EXEMPT 和 IOMBPS_THROTTLE_EXEMPT。如果出现“rsmgr:io rate limit”等待事件，则表示 I/O 达到了限制阈值。

I/O 相关参数的修改语法如下：

```
ALTER SYSTEM SET MAX_IOPS = 1000 SCOPE = BOTH;  --1000 次/s
ALTER SYSTEM SET MAX_MBPS = 200 SCOPE = BOTH;  --200MB/s
```

3.9.3 管理 CDB 资源计划

CDB 资源计划就是分配 PDB 的 share 值和资源限制。以图 3-1 为例进行说明，图 3-1 中显示了默认的 PDB 指令，指定 share 为 1、utilization_limit 为 50%、parallel_server_limit 为 50%。任何 PDB 都是 CDB 的一部分，没有为 PDB 定义的指令，都使用默认的 PDB 指令。图 3-1 中显示了使用默认 PDB 指令的 TESTPDB 和 REPORTPDB。因此，REPORTPDB 和 TESTPDB 各有 1 个 share 且 utilization_ limits 均为 50%。

关于 share 值的含义做进一步的说明，图 3-1 中有 5 个 PDB，根据图示，总的 share 值就是 11（即 3+3+3+1+1），那么对于 CNDBAPDB 这个 PDB，它可以使用 3/11 的 CPU 资源。

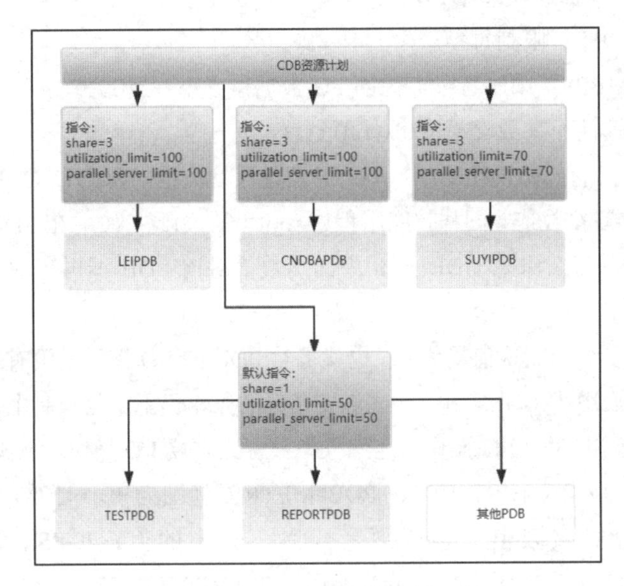

图 3-1 CDB 资源计划

1. 创建 CDB 资源计划

创建 CDB 资源计划的步骤如下：

（1）使用 CREATE_PENDING_AREA 创建等待区域；

（2）使用 CREATE_CDB_PLAN 创建 CDB 资源计划；

（3）使用 CREATE_CDB_PLAN_DIRECTIVE 创建 CDB 资源计划指令；

（4）（可选）使用 UPDATE_CDB_DEFAULT_DIRECTIVE 更新默认的计划

指令；

（5）使用 VALIDATE_PENDING_AREA 使等待区域生效；

（6）使用 SUBMIT_PENDING_AREA 提交等待区域。

在 CDB 资源计划中，关于指令属性的说明如表 3-6 所示。

表 3-6　CDB 资源计划中指令属性的说明

指令属性	说　明
share	CPU 和并行执行服务器资源的资源分配份额
utilization_limit	CPU 资源利用率
parallel_server_limit	PDB 并行执行可以使用的服务器资源的最大百分比。当为一个 PDB 指定了 parallel_server_limit 指令后，该限制的值就是 parallel_server_limit 的值乘以 PARALLEL_SERVERS_TARGET 初始化参数的值。 Oracle 推荐使用 PARALLEL_SERVERS_TARGET 参数而不是 CDB 资源计划中的 parallel_server_limit 指令

创建 CDB 资源计划的具体例子如表 3-7 所示。

表 3-7　创建 CDB 资源计划的例子

PDB	share 指令	utilization_limit 指令	parallel_server_limit 指令
leipdb	2	Unlimited	Unlimited
cndbapdb	2	Unlimited	50%

表 3-7 中为 leipdb 分配 2 个 share，utilization_limit 没有限制，parallel_server_limit 也没有限制；为 cndbapdb 分配 2 个 share，utilization_limit 没有限制，但是 parallel_server_limit 被设置为 50%。

创建表 3-7 所示的 CDB 资源计划的步骤如下。

（1）创建等待区域，代码如下：

```
SQL>exec DBMS_RESOURCE_MANAGER.CREATE_PENDING_AREA();
```

（2）创建 CDB 资源计划。

通过 CREATE_CDB_PLAN 存储过程创建一个名为 zx_cndba 的资源计划，代码如下：

```
SQL>BEGIN
  DBMS_RESOURCE_MANAGER.CREATE_CDB_PLAN(
```

```
    plan    => 'zx_cndba',
    comment => 'CDB resource plan for cndba')；
END;
/
```

（3）创建 CDB 资源计划指令。

首先为leipdb创建指令，并指定share、utilization_limit和parallel_server_limit的值，代码如下（值为 100 表示没有限制）：

```
SQL>BEGIN
  DBMS_RESOURCE_MANAGER.CREATE_CDB_PLAN_DIRECTIVE(
    plan                  => 'zx_cndba',
    pluggable_database    => 'leipdb',
    shares                => 2,
    utilization_limit     => 100,
    parallel_server_limit => 100)；
END;
/
```

再为 cndbapdb 创建指令，并指定 share、utilization_limit 和 parallel_server_limit 的值，代码如下：

```
BEGIN
  DBMS_RESOURCE_MANAGER.CREATE_CDB_PLAN_DIRECTIVE(
    plan                  => 'zx_cndba',
    pluggable_database   => 'cndbapdb',
    shares               => 2,
    utilization_limit    => 100,
    parallel_server_limit => 50)；
END;
/
```

如果想对内存使用进行限制，可以指定 memory_min 和 memory_limit 的值，参数值都是百分比。除此之外，其他的 PDB 将使用默认的 PDB 指令。

（4）（可选）更新默认 CDB 资源计划指令。

如果默认的资源计划指令没有满足要求，则可以修改，代码如下：

```
SQL>BEGIN
  DBMS_RESOURCE_MANAGER.UPDATE_CDB_DEFAULT_DIRECTIVE(
    plan                      => 'zx_cndba',
    new_shares                 => 1,
    new_utilization_limit     => 50,
```

```
    new_parallel_server_limit => 50）;
END;
/
```

（5）使等待区域生效，代码如下：

```
SQL>exec DBMS_RESOURCE_MANAGER.VALIDATE_PENDING_AREA（）;
```

（6）提交等待区域，代码如下：

```
SQL>exec DBMS_RESOURCE_MANAGER.SUBMIT_PENDING_AREA（）;
```

2．修改 CDB 资源计划

修改 CDB 资源计划的步骤和创建 CDB 资源计划的步骤相同，只是调用的存储过程不同。用户可以单独修改某个资源计划指令。修改 CDB 资源计划的步骤如下。

（1）创建等待区域，代码如下：

```
SQL>exec DBMS_RESOURCE_MANAGER.CREATE_PENDING_AREA（）;
```

（2）创建 CDB 资源计划。

通过CREATE_CDB_PLAN存储过程创建一个名为new_cndba的资源计划，代码如下：

```
SQL>BEGIN
  DBMS_RESOURCE_MANAGER.UPDATE_CDB_PLAN（
    plan    => 'new_cndba',
    comment => 'CDB resource plan for new cndba'）;
END;
/
```

（3）创建 CDB 资源计划指令。

首先为leipdb创建指令，并指定share、utilization_limit和parallel_server_limit的值，代码如下：

```
SQL>BEGIN
  DBMS_RESOURCE_MANAGER.UPDATE_CDB_PLAN_DIRECTIVE（
    plan                          => 'new_cndba',
    pluggable_database            => 'leipdb',
    new_shares                    => 2,
    new_utilization_limit         => 100,
    new_parallel_server_limit     => 100）;
END;
```

```
/
```

再为 cndbapdb 创建指令，代码如下：

```
SQL>BEGIN
  DBMS_RESOURCE_MANAGER.UPDATE_CDB_PLAN_DIRECTIVE(
    plan                       => 'new_cndba',
    pluggable_database         => 'cndbapdb',
    new_shares                 => 2,
    new_utilization_limit      => 100,
    new_parallel_server_limit => 50);
END;
/
```

（4）使等待区域生效，代码如下：

```
SQL>exec DBMS_RESOURCE_MANAGER.VALIDATE_PENDING_AREA();
```

（5）提交等待区域，代码如下：

```
SQL>exec DBMS_RESOURCE_MANAGER.SUBMIT_PENDING_AREA();
```

3. 删除 CDB 资源计划

删除 CDB 资源计划的步骤与创建 CDB 资源计划的步骤大致相同，还可以单独删除某个计划指令。

（1）创建等待区域，代码如下：

```
SQL>exec DBMS_RESOURCE_MANAGER.CREATE_PENDING_AREA();
```

（2）删除 CDB 资源计划指令，代码如下：

```
SQL>BEGIN
  DBMS_RESOURCE_MANAGER.DELETE_CDB_PLAN_DIRECTIVE(
    plan               => 'zx_cndba',
    pluggable_database => 'leipdb');
END;
/
```

（3）删除 CDB 资源计划，代码如下：

```
SQL>BEGIN
  DBMS_RESOURCE_MANAGER.DELETE_CDB_PLAN(
    plan => 'zx_cndba');
END;
/
```

（4）使等待区域生效，代码如下：

```
SQL>exec DBMS_RESOURCE_MANAGER.VALIDATE_PENDING_AREA () ;
```

（5）提交等待区域，代码如下：

```
SQL>exec DBMS_RESOURCE_MANAGER.SUBMIT_PENDING_AREA () ;
```

4. 启用 CDB 资源计划

查看当前启用的资源计划，代码如下：

```
SQL> show parameter RESOURCE_MANAGER_PLAN
NAME                           TYPE        VALUE
---------------------- ----------- --------------------
resource_manager_plan      string
```

设置新的资源计划，代码如下：

```
SQL>ALTER SYSTEM SET RESOURCE_MANAGER_PLAN = ' zx_cndba';
```

或者在初始化参数中设置新的资源计划，代码如下：

```
RESOURCE_MANAGER_PLAN = ' zx_cndba'
```

再次验证资源计划，代码如下：

```
SQL> show parameter RESOURCE_MANAGER_PLAN
NAME                            TYPE        VALUE
---------------------- ----------- ---------------------
resource_manager_plan   string     zx_cndba
```

5. 禁用 CDB 资源计划

要禁用资源计划，将初始化参数设置为空就可以了，代码如下：

```
SQL>ALTER SYSTEM SET RESOURCE_MANAGER_PLAN = ' ';
```

或者在初始化参数中设置“RESOURCE_MANAGER_PLAN = ”，但是要重启数据库。

6. 查看 CDB 资源计划

查看新创建的资源计划状态，如果 STATUS（状态）显示 pending（待定的），则说明等待区域没有被提交。

```
SQL> COL PLAN FOR A10
SQL> COL STATUS FOR A15
```

```
SQL> COL COMMENTS FOR A40
SQL> SELECT PLAN, STATUS, COMMENTS FROM DBA_CDB_RSRC_PLANS WHERE
PLAN='ZX_CNDBA';
PLAN        STATUS        COMMENTS
---------- ------------------------------------------
ZX_CNDBA    CDB resource plan for cndba
```

在查看资源计划的详细信息时，由于没有对内存使用进行限制，所以显示为空，也就是不限制，代码如下：

```
SQL>SELECT PLAN,
    PLUGGABLE_DATABASE PDB_NAME,
    UTILIZATION_LIMIT,
    PARALLEL_SERVER_LIMIT,
    MEMORY_MIN,
    MEMORY_LIMIT
    FROM DBA_CDB_RSRC_PLAN_DIRECTIVES
    WHERE PLAN = 'ZX_CNDBA';
PLAN    PDB_NAME    UTILIZATION_LIMIT  PARALLEL_SERVER_LIMIT
MEMORY_MIN MEMORY_LIMIT
--------------- ----------- ------------ ---------- ------------
ZX_CNDBA  ORA$AUTOTASK       90
ZX_CNDBA  LEIPDB
ZX_CNDBA  CNDBAPDB                       50
ZX_CNDBA  ORA$DEFAULT_PDB_DIRECTIVE
```

3.9.4 管理 PDB 资源计划

1. PDB 资源计划

PDB 资源计划和 CDB 资源计划相似，都是为了控制在 PDB 中分配资源。不同之处在于，CDB 资源计划可以控制多个 PDB，而 PDB 资源计划只能控制某个 PDB 的资源分配。

在 CDB 中，PDB 资源计划有以下限制：

- 一个 PDB 资源计划不能有子计划；
- 一个 PDB 资源计划最多可以有 8 个消费组；
- 一个 PDB 资源计划不能有多级调度策略。

如果使用 Non-CDB 创建新的 PDB，而 Non-CDB 中可能包含资源计划，那么这些资源计划可能不符合前面提到的限制。在这种情况下，Oracle 会自动将这些资源计划转换为满足需求的等效 PDB 资源计划。原来的资源计划和指令记录在

DBA_RSRC_PLANS和DBA_RSRC_PLAN_DIRECTIVE视图中，其状态是LEGACY。

如果要创建 PDB 资源计划，那么 CDB 资源计划必须满足表 3-8 所示的要求。如果没有创建新的 CDB 资源计划，那么将会使用默认的 CDB 资源计划。

表 3-8　创建 PDB 资源计划对 CDB 资源计划的要求

资　源	CDB 资源计划要求	当不满足要求时
CPU	必须满足以下其中一种情况： • 必须为 PDB 指定 share 值； • 如果 utilization_limit 的参数值小于 100%，则必须指定该参数	PDB 资源计划中关于 CPU 资源的分配将不会被执行。同样，PDB 资源计划中通过使用 utilization_limit 参数来限制 CPU 资源的指令也不会被执行
Parallel Execution Servers	必须满足以下其中一种情况： • 必须为 PDB 指定 share 值； • 如果 parallel_server_limit 的参数值小于 100%，则必须指定该参数	PDB 资源计划中关于并行执行服务器资源的限制指令将不会被执行。同样，PDB 资源计划中通过使用 parallel_server_limit 参数来限制 CPU 资源的指令也不会被执行。但是，可以通过设置 PARALLEL_SERVERS_TARGET 初始化参数来强制执行并进行限制

CDB 资源计划中可以有多个 PDB 资源计划，属于一对多的关系，如图 3-2 所示。

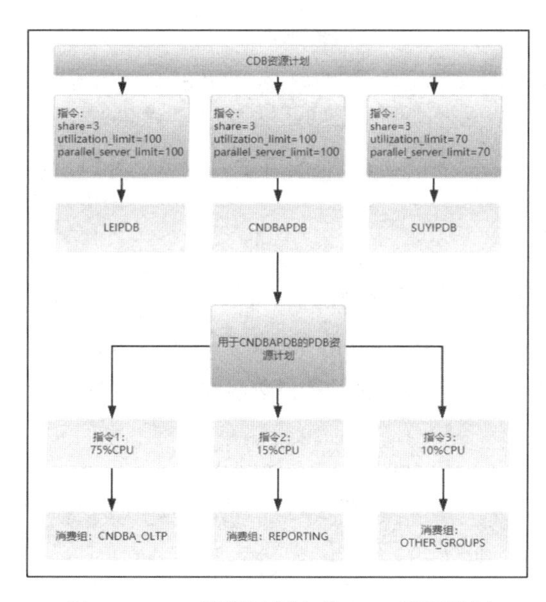

图 3-2　CDB 资源计划中的 PDB 资源计划

从图 3-2 可以看出，实际上在 CDB 资源计划中又进行了资源细分，不同的会话（消费组）使用不同的资源。

2. 创建 PDB 资源计划

创建 PDB 资源计划的步骤如下。

（1）切换到相应 PDB 容器下，代码如下：

```
SQL>ALTER SESSION SET CONTAINER=leipdb;
```

（2）创建等待区域，代码如下：

```
SQL>EXEC DBMS_RESOURCE_MANAGER.CREATE_PENDING_AREA ();
```

如果有等待区域没有被提交，则可以提交或清空等待区域，代码如下：

```
SQL>EXEC DBMS_RESOURCE_MANAGER.CLEAR_PENDING_AREA;
```

（3）创建消费组。

创建一个名为 CNDBA_OLTP 的消费组，管理方法是 ROUND-ROBIN（默认值，意思是所有会话都是公平的），代码如下。还有一个方法是 RUN-TO-COMPLETION，该方法会把长时间运行的会话放到其他会话前面执行。

```
SQL>BEGIN
DBMS_RESOURCE_MANAGER.CREATE_CONSUMER_GROUP (
CONSUMER_GROUP => 'CNDBA_OLTP',
COMMENT        => 'CNDBA OLTP Applications',
MGMT_MTH       => 'ROUND-ROBIN');
END;
/
```

（4）映射会话到相应的消费组中。

ATTRIBUTE 用于指定会话的类型，比如数据库用户的属性是 ORACLE_USER，将用户 LEI 映射到消费组 CNDBA_OLTP 中的语法如下：

```
SQL>BEGIN
DBMS_RESOURCE_MANAGER.SET_CONSUMER_GROUP_MAPPING (
    ATTRIBUTE      => DBMS_RESOURCE_MANAGER.ORACLE_USER,
    VALUE          => 'LEI',
    CONSUMER_GROUP => 'CNDBA_OLTP');
END;
/
```

查看各个属性的优先级，数字 1 表示优先级最高，数字越大优先级越低，代码如下：

```
SQL> select * from DBA_RSRC_MAPPING_PRIORITY;
ATTRIBUTE                         PRIORITY STATUS
------------------------------ ---------- ----------
EXPLICIT                                1
SERVICE_MODULE_ACTION               2
SERVICE_MODULE                          3
MODULE_NAME_ACTION                  4
MODULE_NAME                         5
SERVICE_NAME                        6
ORACLE_USER                             7
CLIENT_PROGRAM                          8
CLIENT_OS_USER                          9
CLIENT_MACHINE                          10
CLIENT_ID                           11
```

（5）创建资源计划。调用 DBMS_RESOURCE_MANAGER.CREATE_PLAN 存储过程创建资源计划，其中计划名称和注释是必须有的，其他参数都是可选的，代码如下：

```
SQL>BEGIN
  DBMS_RESOURCE_MANAGER.CREATE_PLAN (
   PLAN    => 'DAYTIME',
   COMMENT => 'More resources for CNDBA OLTP applications') ;
END;
/
```

（6）创建资源计划指令。

可以通过 CREATE_PLAN_DIRECTIVE 命令创建资源计划指令，该存储过程的参数很多，具体可以参考官方手册。下面列举几个参数。

- SWITCH_GROUP：默认值是 null，表示不指定转换组。数据库预定义的值有 CANCEL_SQL、KILL_SESSION、LOG_ONLY，创建的新消费组名称不能和这三个名称相同。当满足切换条件后，就会调用指定的消费组，例如 SWITCH_GROUP= CANCEL_SQL，当前调用将会被取消。
- UTILIZATION_LIMIT：资源限制，但不会限制 Parallel Servers。
- MGMT_Pn：*n* 的取值为 1~8，在第 *n* 级别指定资源分配的百分比，代替了 cpu_pn 参数。

创建资源计划指令的代码如下：

```
SQL>BEGIN
  DBMS_RESOURCE_MANAGER.CREATE_PLAN_DIRECTIVE
    (PLAN              => 'DAYTIME',
    GROUP_OR_SUBPLAN => 'CNDBA_OLTP',
    COMMENT           => 'CNDBA OLTP group',
    MGMT_P1           => 75）;--75%的CPU使用百分比
  DBMS_RESOURCE_MANAGER.CREATE_PLAN_DIRECTIVE
    (PLAN              => 'DAYTIME',
    GROUP_OR_SUBPLAN => 'OTHER_GROUPS', --必须指定OTHER_GROUPS
    COMMENT           => 'Lowest priority sessions',
    MGMT_P1           => 1）;
END;
/
```

（7）验证等待区域，代码如下：

```
SQL>EXEC DBMS_RESOURCE_MANAGER.VALIDATE_PENDING_AREA();
```

（8）提交等待区域，代码如下：

```
SQL>EXEC DBMS_RESOURCE_MANAGER.SUBMIT_PENDING_AREA();
```

3. 启用 PDB 资源计划

启动 PDB 资源计划，需要切换到相应的 PDB 容器下，执行以下代码：

```
SQL>ALTER SYSTEM SET RESOURCE_MANAGER_PLAN = 'daytime';
```

4. 修改 PDB 资源计划

修改 PDB 资源计划的步骤和修改 CDB 资源计划的步骤相似，可以通过调用 DBMS_RESOURCE_MANAGER 包来修改 PDB 资源计划。

（1）创建等待区域，代码如下：

```
SQL>EXEC DBMS_RESOURCE_MANAGER.CREATE_PENDING_AREA();
```

（2）修改资源计划。

- 调用 UPDATE_CONSUMER_GROUP 可以修改消费组。
- 调用 DELETE_CONSUMER_GROUP 可以删除消费组。
- 调用 UPDATE_PLAN 可以更新资源计划。
- 调用 DELETE_PLAN 可以删除资源计划。
- 调用 UPDATE_PLAN_DIRECTIVE 可以更新资源计划指令。

- 调用 DELETE_PLAN_DIRECTIVE 可以删除资源计划指令。

（3）验证等待区域，代码如下：

```
SQL>EXEC DBMS_RESOURCE_MANAGER.VALIDATE_PENDING_AREA();
```

（4）提交等待区域，代码如下：

```
SQL>EXEC DBMS_RESOURCE_MANAGER.SUBMIT_PENDING_AREA();
```

5. 禁用 PDB 资源计划

禁用 PDB 资源计划，需要切换到相应的 PDB 容器下，执行以下代码：

```
SQL>ALTER SYSTEM SET RESOURCE_MANAGER_PLAN = '';
```

3.9.5　监控 PDB

Oracle 提供了一组用于监控 PDB 资源管理器的动态性能视图。

1. 用于监控 PDB 资源管理器的视图

- V$RSRCPDBMETRIC：该视图提供了 PDB 当前消耗的资源统计信息，包括 CPU、I/O 和内存等。
- V$RSRCPDBMETRIC_HISTORY：和 V$RSRCPDBMETRIC 视图相似，唯一的区别就是 V$RSRCPDBMETRIC 视图只记录过去 1 分钟的统计信息，而 V$RSRCPDBMETRIC_HISTORY 视图记录过去 60 分钟的统计信息。
- V$RSRC_PDB：该视图用于记录资源计划启用以来所有的统计信息。
- DBA_HIST_RSRC_PDB_METRIC：该视图包含 V$RSRCPDBMETRIC_HISTORY 视图的内容，用于 AWR 报告。

注意，当把 STATISTICS_LEVEL 初始化参数设置为 ALL 或 TYPICAL 时，V$RSRCPDBMETRIC 视图和 V$RSRCPDBMETRIC_HISTORY 视图会记录当前还没有被资源管理器管理的资源的统计信息。

2. 监控 PDB 的 CPU 使用情况

通过 V$RSRCPDBMETRIC 视图，可以根据过去 1 分钟的会话数或利用率来跟踪 CPU 的使用情况（以 ms 为单位）。该视图提供的是实时数据，对于观察 PDB 的 CPU 使用情况非常有用。

也可以通过该视图来比较 PDB 之间的 CPU 使用情况，例如：

- 使用的 CPU 时间；
- CPU 等待时间；
- 消耗 CPU 的平均会话数；
- 等待分配 CPU 资源的会话数。

（1）根据 PDB 的 CPU 利用率跟踪 CPU 的使用情况

根据 PDB 的 CPU 利用率跟踪 CPU 的使用情况，需要查询 V$RSRCPDBMETRIC 视图中的 CPU_UTILIZATION_LIMIT 列和 AVG_CPU_UTILIZATION 列。AVG_CPU_UTILIZATION 列表示 PDB 所消耗的 CPU 的平均百分比，CPU_UTILIZATION_LIMIT 列表示 PDB 可以使用的 CPU 的最大百分比。这个操作是使用 UTILIZATION_LIMIT 指令来设置的，代码如下：

```
SQL>COLUMN PDB_NAME FORMAT A10
SQL>SELECT r.CON_ID,
     p.PDB_NAME,
     r.CPU_UTILIZATION_LIMIT,
     r.AVG_CPU_UTILIZATION
FROM  V$RSRCPDBMETRIC r,
     CDB_PDBS p
WHERE r.CON_ID = p.CON_ID;
 CON_ID PDB_NAME  CPU_UTILIZATION_LIMIT  AVG_CPU_UTILIZATION
---------- ------------------------------- -------------------
  3   LEIPDB                     100                0
  4   CNDBAPDB                    100                0
```

（2）跟踪 PDB 的 CPU 使用情况和等待时间

通过 V$RSRCPDBMETRIC 视图中的 CPU_CONSUMED_TIME 列和 CPU_TIME_WAIT 列可以跟踪每个 PDB 的 CPU 使用情况和等待时间（以 ms 为单位），NUM_CPUS 列表示资源管理器正在管理的 CPU 数量，代码如下：

```
SQL>COLUMN PDB_NAME FORMAT A10
SQL>SELECT r.CON_ID,
     p.PDB_NAME,
     r.CPU_CONSUMED_TIME,
     r.CPU_WAIT_TIME,
     r.NUM_CPUS
FROM  V$RSRCPDBMETRIC r,
```

```
      CDB_PDBS p
WHERE r.CON_ID = p.CON_ID;
 CON_ID  PDB_NAME   CPU_CONSUMED_TIME  CPU_WAIT_TIME   NUM_CPUS
---------- ------------------ ---------------- ----------------
  3    LEIPDB                   0            0          1
  4    CNDBAPDB                 0            0          1
```

（3）根据 PDB 的会话数查看 CPU 的使用情况和等待时间

根据 PDB 的会话数查看 CPU 的使用情况和等待时间，需要使用 V$RSRCPDBMETRIC 视图中的 RUNNING_SESSIONS_LIMIT、AVG_RUNNING_SESSIONS 和 AVG_WAITING_SESSIONS 三列。

- RUNNING_SESSIONS_LIMIT 列表示在指定的 PDB 中可以运行的最大会话数。这个操作是由 UTILIZATION_LIMIT 指令设置的。
- AVG_RUNNING_SESSIONS 列表示正在使用 CPU 的平均会话数量。
- AVG_WAITING_SESSIONS 列表示正在等待 CPU 的平均会话数量。

根据 PDB 的会话数查看 CPU 的使用情况和等待时间的代码如下：

```
    SQL>COLUMN PDB_NAME FORMAT A10
    SQL>SELECT r.CON_ID, p.PDB_NAME, r.RUNNING_SESSIONS_LIMIT,
r.AVG_RUNNING_SESSIONS,    r.AVG_WAITING_SESSIONS FROM
V$RSRCPDBMETRIC r, CDB_PDBS p WHERE r.CON_ID = p.CON_ID;
    CON_ID PDB_NAME   RUNNING_SESSIONS_LIMIT AVG_RUNNING_SESSIONS
AVG_WAITING_SESSIONS
    ---------------------- -------------- ----------------------
    -------------------------------- ---------------------------
     3     LEIPDB                    1                0           0
     4     CNDBAPDB                  1                0           0
```

3. 监控 PDB 的并行执行（Parallel Execution）情况

可以通过 V$RSRCPDBMETRIC 视图来跟踪 PDB 的并行语句和并行服务器的使用。要跟踪 PDB 的并行语句和并行服务器（Parallel Server）的使用，需要使用该视图中的 AVG_ACTIVE_PARALLEL_STMTS、AVG_QUEUED_PARALLEL_STMTS、AVG_ACTIVE_PARALLEL_SERVERS、AVG_QUEUED_PARALLEL_SERVERS 和 PARALLEL_SERVERS_LIMIT 五列。

- AVG_ACTIVE_PARALLEL_STMTS 列和 AVG_ACTIVE_PARALLEL_SERVERS 列表示并行语句运行的平均数和并行语句使用的平均并行服务器数。

- AVG_QUEUED_PARALLEL_STMTS 列和 AVG_QUEUED_PARALLEL_SERVERS 列表示排队的并行语句请求的平均并行服务器数和平均并行服务器数。
- PARALLEL_SERVERS_LIMIT 列表示 PDB 允许使用的并行服务器数。

监控 PDB 的并行执行情况的代码如下：

```
SQL>COLUMN PDB_NAME FORMAT A10
SQL>SELECT r.CON_ID, p.PDB_NAME, r.AVG_ACTIVE_PARALLEL_STMTS,
r.AVG_QUEUED_PARALLEL_STMTS, r.AVG_ACTIVE_PARALLEL_SERVERS,
r.AVG_QUEUED_PARALLEL_SERVERS, r.PARALLEL_SERVERS_LIMIT FROM
V$RSRCPDBMETRIC r, CDB_PDBS p WHERE r.CON_ID = p.CON_ID;
CON_ID PDB_NAME   AVG_ACTIVE_PARALLEL_STMTS
AVG_QUEUED_PARALLEL_STMTS AVG_ACTIVE_PARALLEL_SERVERS
AVG_QUEUED_PARALLEL_SERVERS PARALLEL_SERVERS_LIMIT
---------------------- -------------- ----------------------
---------------------- -------------- ----------------------
 3 LEIPDB          0       0          0          0          20
4 CNDBAPDB        0       0        0           0          10
```

4．监控 PDB 产生的 I/O 情况

可以使用 V$RSRCPDBMETRIC 视图监控 PDB 产生的 I/O 情况。通过设置 PDB 中的 MAX_IOPS 或 MAX_MBPS 初始化参数可以限制 PDB 产生的 I/O。使用 V$RSRCPDBMETRIC 视图比较 PDB 之间产成的 I/O，按照每秒的读写次数和每秒的数据量（以 MB 为单位）进行比较。

（1）跟踪 PDB 每秒产生的 I/O 读写次数

跟踪 PDB 每秒产生的 I/O 读写次数的代码如下：

```
SQL>COLUMN PDB_NAME FORMAT A10
SQL>SELECT r.CON_ID, p.PDB_NAME, r.IOPS FROM V$RSRCPDBMETRIC r,
CDB_PDBS p WHERE r.CON_ID = p.CON_ID;
CON_ID  PDB_NAME     IOPS
---------- --------------- ------------------
  3    LEIPDB    .216522318
  4    CNDBAPDB   .216522318
```

（2）跟踪 PDB 每秒产生的 I/O 数据量

跟踪 PDB 每秒产生的 I/O 数据量的代码如下：

```
SQL>COLUMN PDB_NAME FORMAT A10
SQL>SELECT r.CON_ID, p.PDB_NAME, r.IOMBPS FROM V$RSRCPDBMETRIC r,
CDB_PDBS p WHERE r.CON_ID = p.CON_ID;
CON_ID PDB_NAME      IOMBPS
---------- ---------- ----------
  3    LEIPDB           0
  4    CNDBAPDB          0
```

5. 监控 PDB 的内存使用情况

通过 V$RSRCPDBMETRIC 视图可以跟踪 PDB 的内存使用情况。可以查看 PDB 的 SGA、 PGA、Buffer Cache（缓存）和 Shared Pool Memory（共享池内存）的使用情况。通过该视图的 SGA_BYTES、PGA_BYTES、BUFFER_CACHE_BYTES 和 SHARED_POOL_BYTES 四列可以分别查看相应内存区域的使用情况。

监控 PDB 的内存使用情况的代码如下（数字均以字节为单位）:

```
SQL>COLUMN PDB_NAME FORMAT A10
SQL>SELECT r.CON_ID, p.PDB_NAME, r.SGA_BYTES, r.PGA_BYTES,
r.BUFFER_CACHE_BYTES, r.SHARED_POOL_BYTES FROM V$RSRCPDBMETRIC r,
CDB_PDBS p WHERE r.CON_ID = p.CON_ID;
      CON_ID PDB_NAME    SGA_BYTES  PGA_BYTES BUFFER_CACHE_BYTES
SHARED_POOL_BYTES
----- ---------- ---------- ------------------ -----------------
        3 LEIPDB     81596934  10463259     61335142      20261792
        4 CNDBAPDB   95215664   4952419     80732160      14483504
```

第4章

In-Memory 概念

In-Memory（IM）是 Oracle 12c 中引入的新特性，可以极大地提高实时分析查询的性能。在 IM 特性中最重要的是列式存储（IM Column Store）。

一般情况下，关系型数据库以行或列的格式来存储数据。在内存和磁盘中，存储数据的格式是相同的。Oracle 将行连续存储在数据块中。例如，在一个有三行数据的表中，数据块首先存储第一行数据，然后存储第二行数据，最后存储第三行数据。每行包含该行的所有列值。以行格式存储的数据针对事务处理进行了优化。例如，更新少量行中的所有列的列值只需要修改少量的块。

但是以行格式存储数据对于分析查询来说有一个很大的缺点，即通常情况下分析查询只需要查询表中某几个字段而不是所有字段。列式存储在这方面有了很大的性能提升，列式存储以表的每列进行存储，列与列之间是分开存储的。

分析查询只需要访问少量列，却扫描了整个表。因此，列式存储对分析来说效率是最高的。由于列是分开存储的，因此分析查询只需要访问所需的列，从而避免了读取不必要的数据。

由于目前大部分的数据库厂商只能同时提供一种存储方式，要么以列式存储，要么以行式存储，所以 Oracle 在 12c 中引入了这个关键的特性，用户可以根据自己的实际需求来选择合适的数据存储方式。

4.1 Oracle IM 解决方案

Oracle IM 包括 IM 列式存储、高级查询优化和可用性解决方案。IM 可以为数据仓库和混合使用的数据库分析查询提供更快的查询速度。

4.1.1 IM 列式存储

IM 列式存储以特殊的压缩列格式来保存表、分区和单个列的副本，该列式存储针对快速扫描进行了专门优化。

IM 列式存储在 SGA 中的一个可选的内存区（In-Memory Area）中。IM 列存储不会替换基于行的存储或数据库缓冲区的缓存，相反会对其进行补充。数据库使数据能够以行和列的格式存储在内存中，提供了两全其美的解决方案。IM 列式存储保存了一份独立于磁盘格式的表数据的完整复制，保证事务的一致性。同时，IM 列式存储中的数据不需要再次加载到高速缓冲区中。

在以下数据库级别上的 DDL 语句中指定 IM 参数就可以启用 IM 列式存储：

- 列（虚拟、非虚拟）；
- 表（内部表、外部表）、物化视图、分区；
- 表空间。

默认情况下，如果表空间启用了 IM 列式存储，那么该表空间中的所有新创建的表、物化视图都会启动 IM 列式存储。IM 在填充数据时，会将磁盘上基于行的数据自动转换为 IM 列式存储中的列数据。可以在 IM 列式存储中配置数据库对象列的全部或部分列填充。同样，对于分区表或物化视图，也可以配置全部或部分分区来填充。

这里的填充是从数据文件中读取现有数据块，将行格式转换为列格式，然后将列数据写入 IM 列式存储。加载指使用 DML 或 DDL 语句将新的数据插到数据库中。

4.1.2 高级查询优化

IM 除列式存储外，还包括一些对分析查询的性能优化。

- 表达式由一个或多个值、运算符、解析为一个值的 SQL 函数（仅限于 DETERMINISTIC）组合而来。默认情况下，IM 表达式通过 DBMS_INMEMORY_ADMIN.IME_CAPTURE_EXPRESSIONS 存储过程识别“热（使用率高的）”表达式并填充到 IM 列式存储中。IM 表达式作为隐藏的虚拟列，其访问方式和非虚拟列的访问方式相同。

- 连接组（Join Group）是一个用户定义的对象，用于指定常用于连接一组表的列。在某些查询中，连接组可以使数据库避免解压缩和散列列值的性能开销。
- 对于将数据量小的维度表连接到庞大的事实表的聚合查询，IM 聚合使用 VECTOR GROUP BY 方式来提高性能。这种方式会在扫描大表的过程中汇总数据，而不是在扫描后再进行数据汇总。
- 在 IM 列式存储中，当 IMCU 中的数据发生显著变化时，重新填充会自动更新 IMCU 中的数据。如果 IMCU 中有陈旧的数据，但对应的数量又没有达到过期的阈值，那么后台进程就可能采用渐进的方式对 IMCU 进行重新填充操作，也就是每次只选择一部分满足重新加载条件的 IMCU 进行处理，具体可以通过 INMEMORY_TRICKLE_ REPOPULATE_ SERVERS_PERCENT 参数进行调整。

本小节涉及一些专业术语，简单解释如下。

- IMCU（In-Memory Compression Unit）: IM 列式存储中的存储单元经过优化以实现更快的扫描，它将表中的每列分开存储并进行压缩。IMCU 和一组数据库块之间是一对多的关系。例如，如果一个表包含 c1 列和 c2 列，并且它的行存储在磁盘上的 100 个数据库块中，则 IMCU 1 可能会存储 1~50 数据库块的这两列值、IMCU 2 可能会存储 51~100 数据库块的这两列值。
- 重新填充（Repopulation）：IMCU 中的数据被修改后，会自动刷新当前 IMCU 中的数据。
- 基于阈值的重新填充（Threshold-based Repopulation）：当 IMCU 中陈旧数据的数量达到定义的阈值时，会自动重新填充 IMCU。IMCU 后台进程会每隔一段时间（默认为 2 分钟）检查一次是否有 IMCU 满足重新加载的条件，如果发现了满足条件的 IMCU，就会通知工作进程对相应的 IMCU 进行重新填充。
- 渐进的重新填充（Trickle Repopulation）：它是基于阈值的重新填充的补充。由于重新填充的成本比较高，可能会影响一些正在运行的语句，所以当 IM 列式存储中陈旧数据的数量还没有达到过期阈值时，Oracle 就会采用渐进的方式对 IMCU 进行重新填充。

4.1.3　支持高可用

这里所说的可用性是指应用程序、服务或功能按需访问的程度。Oracle IM 支持以下可用特性。

- IM FastStart 可以显著减少在数据库实例重启后重新载入数据到 IM 列式存储中的时间。IM FastStart 是通过定期保存当前 IM 列式存储中的数据（以压缩的列式格式）的副本到磁盘上来实现的。
- Oracle RAC 每个节点都有自己的 IM 列式存储。可以在每个节点上存储不同的对象（也可以存储相同的对象），或者一个大的对象可以分开存储在各个节点上。
- 从 Oracle 18c 开始，在 ADG 环境下备库也支持 IM 列式存储。

4.1.4　提高分析查询的性能

IM 压缩的列格式可实现更快的扫描、查询、连接和聚合。

1. 提高扫描的性能

IM 列式存储可以为扫描大量数据的查询提供更大的吞吐量。IM 列式存储能够实时分析数据，特别是可以显著提高以下操作的查询性能：

- 扫描大量行并使用了如<、>、=和 IN 等运算符的操作；
- 从表或具有多列的物化视图中选择某几列的操作，例如访问一个有 100 列的表中的 5 列；
- 使用 SQL 运算符查询 LOB 列的操作。

列格式式对大多数数字和短字符串数据类型使用固定宽度的列，这样可实现快速的矢量处理，从而使数据库能够更快地响应查询。

IM 列式存储的扫描比普通行存储的扫描要快的主要原因有以下几点。

- 减少高速缓冲区的开销：IM 列式存储以列格式存储数据，数据不会保存在数据文件中或产生 redo 日志，因此就避免了从磁盘读取数据到高速缓冲区中的开销。
- 数据修剪：Oracle 只扫描需要的列，比扫描整行数据要快得多。此外，数据库使用存储索引和内部数据字典只读取特定查询所需的 IMCU。例如，如果只需要查询成绩高于 90 分的学生的姓名，那么 Oracle 会通过 IMCU 修剪（Pruning）来去除不满足条件的数据。

- 压缩：正常情况下，压缩是为了节省空间，但是在 IM 列式存储中，压缩是为了更快地扫描。Oracle 可以自动在 WHERE 子句的谓词上使用压缩算法压缩列存储的数据。根据所应用的压缩类型，Oracle 数据库可以以其压缩格式扫描数据，而不用先解压数据。因此，数据库在 IM 列式存储中扫描的数据量必须小于数据库缓冲区高速缓存中的数据量。
- 矢量处理（Vector Processing）：每个 CPU 内核扫描本地的内存列。要以数组（集合）的形式处理数据，扫描需要使用 SIMD 矢量指令。例如，查询可以在单个 CPU 指令中读取一组值，而不是逐个去读取。CPU 内核的矢量扫描速度比行扫描速度要快几个数量级。

2．提高连接的性能

Bloom Filter 是测试集合中对象的低内存（Low-Memory）数据结构。IM 列式存储使用 Bloom Filter 来提高连接的性能。

Bloom Filter 通过将小维度表上的谓词部分转换为大的事实表（Fact Table）上的过滤器来加速连接操作。这种优化对多个维度表和一个大事实表进行连接的操作很有用。如果事实表上的键值有很多重复，那么连接速度可以加快几个数量级。

3．提高聚合的性能

分析查询的一个重要方面就是通过汇总数据来查看数据的分布和趋势。当数据存储在 IM 列式存储中时，聚合和复杂的 SQL 查询运行得更快。

在 Oracle 12c 之前，聚合都是通过 GROUP BY 方式来实现的。从 Oracle 12.1 开始，提供了 VECTOR GROUP BY 方式来实现更快的 IM，进行基于数组的聚合操作。在扫描事实表期间，数据库将聚合值累加到内存数组中，并使用高效算法执行聚合。基于主键和外键关系的连接，已针对星型模式和雪花模式进行了优化。

4．总结

IM 列式存储对 OLTP 系统的性能提升并不是直接体现的。例如，对于本来需要创建更多索引来加快查询速度的表，如果将表存储到 IM 中，则不需要创建那么多的索引就能够达到更快的查询速度，而更少的索引对表的插入操作性能也会有所提升。

对于以下查询，IM 无法有效地提升查询性能：

- 复杂条件的查询；
- 查看太多列的查询；
- 返回太多行的查询。

4.1.5 启用 IM 的条件

启用 IM 的条件只有两个：

- 内存最小是 100MB；
- COMPATIBLE 初始化参数值一定是 12.1.0 及以上。

启用 IM 特性后，如果操作中出现“TABLE ACCESS IN MEMROY FULL”提示，则表示其中部分或全部数据是从 IM 中查询的。

4.2 IM 列式存储架构

IM 列式存储使用列格式将表和分区存储在内存中用于快速查询。Oracle 数据库在使用复杂体系结构的同时管理列格式和行格式存储的数据。

若启用了 IM 列式存储，那么 SGA 将同时管理 In-Memory 内存区域和数据库高速缓冲区中的数据。

IM 列式存储用列存储格式来编码数据，每个列都是独立的结构。这些列是连续存储的，可以优化分析查询。数据库缓冲区中的高速缓存可以修改存储在 IM 列式存储中的对象，但缓冲区中的缓存还是用传统行格式来存储数据的，连续数据块用于存储行，可以提供更好的事务性能。

1. IM 内存区域

IM 内存区域是由 INMEMORY_SIZE 初始化参数控制的，默认情况下的参数值是 0，也就意味着没有启用 IM。如果要启用 IM，则需要分配最小 100MB（INMEMORY_SIZE=100MB）的内存，可以通过 V$SGA 视图和 V$INMEMORY_ AREA 视图查看。

IM 内存区域的大小将从 SGA_TARGET 初始化参数值中减去，如图 4-1 所示。例如，如果将 SGA_TARGET 设置为 10 GB，并且将 INMEMORY_SIZE 设置为 4 GB，则将 40%的 SGA_TARGET 大小分配给 IM 内存区域。

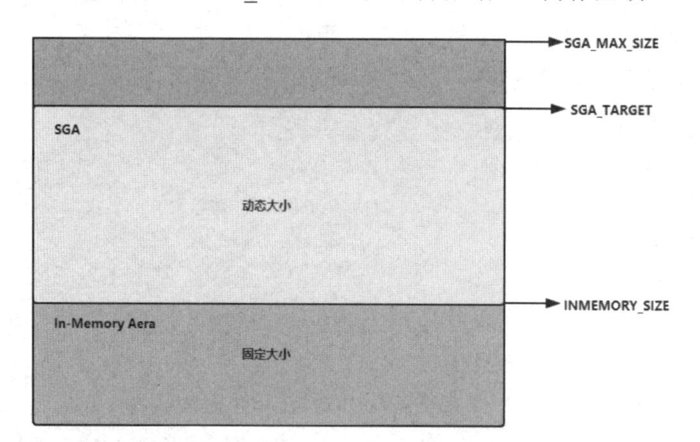

图 4-1 In-Memroy 内存区域

和 SGA 的其他内存组成部分不同，IM 无法通过自动内存管理来进行管理。但是从 Oracle 12.2 开始，INMEMROY_SIZE 支持动态修改（无法减小），但是要满足以下条件：

- SGA 还有空闲内存；
- 新的 INMEMORY_SIZE 的大小至少要比当前 INMEMROY_SIZE 的大小大 128MB。

2. IM 内存区域中的内存池

IM 内存区域被分为用于存储列数据的内存池和用于存储元数据的内存池。

- 列数据池：用于存储包含列数据的 IMCU，V$INMEMORY_AREA 视图中的 POOL 列将该数据池标识为 1MB POOL。
- 元数据池：用于存储有关存储在 IM 列式存储中的对象的元数据，V$INMEMORY_AREA 视图中的 POOL 列将该数据池标识为 64KB POOL。

注意，这两个内存池的大小无法手动修改。

查看内存池的大小，代码如下：

```
SQL>COL POOL FORMAT a9
SQL>COL POPULATE_STATUS FORMAT a15
SQL>SELECT CON_ID,POOL, TRUNC (ALLOC_BYTES/ (1024*1024*1024) ,2)
"ALLOC_GB",
```

```
        TRUNC (USED_BYTES/ (1024*1024*1024) ,2)  "USED_GB",
        POPULATE_STATUS
FROM    V$INMEMORY_AREA;

   CON_ID POOL       ALLOC_GB   USED_GB POPULATE_STATUS
---------- --------- ---------- ---------- ---------------
     1 1MB POOL      0     0 OUT OF MEMORY
     1 64KB POOL     0     0 OUT OF MEMORY
     2 1MB POOL      0     0 OUT OF MEMORY
     2 64KB POOL     0     0 OUT OF MEMORY
     3 1MB POOL      0     0 OUT OF MEMORY
     3 64KB POOL     0     0 OUT OF MEMORY
```

查看当前 IM 内存的大小，代码如下：

```
SQL>SELECT NAME, VALUE/ (1024*1024*1024)  "SIZE_IN_GB"  FROM
V$SGA  WHERE  NAME LIKE '%Mem%';
NAME         SIZE_IN_GB
-------------------- ----------
In-Memory Area      10
```

第 5 章

配置 In-Memory 列式存储

5.1 启用 IM 列式存储并指定大小

启用 IM 列式存储只要设置 INMEMORY_SIZE 参数即可。默认情况下，INMEMORY_SIZE 为 0，即没有启用 IM 列式存储。将 INMEMORY_SIZE 设置为非 0，然后重启数据库就可以启用 IM 列式存储了，启用后 INMEMORY_SIZE 支持动态修改大小。

默认情况下，必须使用表、表空间或物化视图的 CREATE 或 ALTER 语句的 INMEMORY 子句，将指定对象存储到 IM 列式存储中。

5.1.1 预估 IM 列式存储所需的大小

根据实际需要估算 IM 列式存储的大小，然后动态调整 IM 列式存储的大小以满足这些要求。可以使用压缩来减少占用的内存大小。

IM 列式存储所需的内存大小取决于存储在其中的数据库对象及每个对象使用的压缩方法。在为数据库对象选择压缩方法时，需要平衡性能和内存大小。

- 如果想节省内存，则选择 FOR CAPACITY HIGH 或 FOR CAPACITY LOW 压缩方法，但是在查询数据时，会消耗更多的 CPU 资源来解压缩。
- 如果想获得更高的性能，则选择 FOR QUERY HIGH 或 FOR QUERY LOW 压缩方法，但是会占用更多的内存。

指定 IM 列式存储的大小，可以参考以下准则。

- 对于要填充到 IM 列式存储中的每个对象，估计其消耗的内存量。Oracle 压缩顾问使用 MEMCOMPRESS 子句估算压缩比率，是使用

DBMS_COMPRESSION 过程来实现的。

- 将所有对象的大小累加起来，要注意在对象存储到 IM 列式存储中后，V$IM_SEGMENTS 视图中将显示对象在磁盘上的实际大小及它们在 IM 列式存储中的大小。可以根据信息来计算对象的压缩比率。但是，如果对象在磁盘上被压缩，则此查询不会显示正确的压缩比率。
- 如果配置了 IM 优化算法，并且 IM 表使用 FOR QUERY LOW 方法进行压缩，那么需要增加大约 15%的空间来解决 NUMBER 列的双重存储问题。

5.1.2　启用 IM 列式存储的具体步骤

查看当前 INMEMROY_SIZE 的大小，代码如下：

```
SQL> show con_name
CON_NAME
------------------------------
CDB$ROOT
SQL> set lines 120
SQL> SHOW PARAMETER INMEMORY_SIZE
NAME                             TYPE        VALUE
------------------------------------ ----------- -----
inmemory_size                    big integer 0
```

修改 INMEMROY_SIZE 的大小，INMEMROY_SIZE 最小为 100MB，代码如下：

```
SQL> ALTER SYSTEM SET INMEMORY_SIZE=100M SCOPE=SPFILE;
System altered.
SQL> shutdown immediate
Database closed.
Database dismounted.
ORACLE instance shut down.
SQL> startup
ORACLE instance started.

Total System Global Area 1191181104 bytes
Fixed Size           8895280 bytes
Variable Size          922746880 bytes
Database Buffers   134217728 bytes
Redo Buffers         7880704 bytes
```

```
In-Memory Area         117440512 bytes   #注意这里多了一个 IM 区域
Database mounted.
Database opened.
SQL> show parameter inmemory_size
NAME                TYPE             VALUE
------------------ --------------- --------------------
inmemory_size       big integer     112M
```

5.1.3 动态修改 IM 列式存储的大小

从 Oracle 12.2 开始支持动态修改 INMEMORY_SIZE 的大小，但是只能动态增加，无法动态减少。如果想要减少 INMEMORY_SIZE 的大小，那么必须指定 SCOPE=SPFILE 并重启数据库实例，使其生效。

动态增加 INMEMORY_SIZE 的大小，必须满足以下条件：

- 已启用 IM 列式存储；
- COMPATIBLE 初始化参数必须设置为 12.2.0 或更高；
- 数据库必须以 SPFILE 参数文件的方式打开；
- INMEMORY_SIZE 每次增加的大小至少要比当前 INMEMORY_SIZE 的大小大 128Mb；
- 以 SCOPE=BOTH 或 SCOPE=MEMROY 方式修改，否则需要重启数据库实例才能生效。

修改 IM 列式存储大小的示例代码如下：

```
SQL> show parameter inmemory_size
NAME                       TYPE             VALUE
-------------------------------- -----------------------
inmemory_size              big integer     112M
SQL> alter system set inmemory_size = 221m scope=both;
alter system set inmemory_size = 221m scope=both
*
ERROR at line 1:
ORA-02097: parameter cannot be modified because specified value
is invalid
ORA-02095: specified initialization parameter cannot be modified
SQL> alter system set inmemory_size = 250m scope=both;
System altered.
SQL> show parameter inmemory_size
NAME                           TYPE             VALUE
---------------------------- ----------------------
```

```
inmemory_size                    big integer     250M
```

注意，如果增加的大小不能满足要求，就会报 ORA-02097 和 ORA-02095 错误，这里并不是不能改，而是不满足条件。

5.2　禁用 IM 列式存储

要禁用 IM 列式存储，需要将 INMEMORY_SIZE 设置为 0（SPFILE），然后重启数据库实例即可。

5.3　将对象存储到 IM 列式存储中

默认情况下，Oracle 不会自动将对象存储到 IM 列式存储中，需要用户手动指定对象。另外需要注意的是，如果一个 segment（分段）在磁盘上占用的空间小于 64KB，那么将不会被存储到 IM 列式存储中。

注意以下对象不支持 IM 列式存储：

- 索引；
- 索引组织表；
- 散列群集（Hash Cluster）；
- SYS 用户的对象和存储在 SYSTEM 或 SYSAUX 表空间中的对象。

将对象存储到 IM 列式存储中的示例代码如下：

```
SQL> show pdbs

   CON_ID CON_NAME              OPEN MODE  RESTRICTED
---------- ------------------------------ ---------- ----------
     2 PDB$SEED            READ ONLY  NO
     3 DAVE            READ WRITE NO
SQL> alter session set container=dave;
SQL> create user cndba identified by cndba;
User created.
SQL> grant connect,resource,dba to cndba;
Grant succeeded.
SQL> create table cndba.cndba INMEMORY PRIORITY LOW as select *
from dba_objects;
Table created.
SQL> select segment_name, populate_status from  v$im_segments
where segment_name = 'CNDBA';
no rows selected
SQL> select count (1) from cndba.cndba; #查询表，已经将数据加载
```

```
到内存
      COUNT（1）
    ----------
         72918
    SQL> select segment_name, populate_status from  v$im_segments
where  segment_name = 'CNDBA';
    SEGMENT_NAME     POPULATE_STATUS
    ----------------- ---------------
    CNDBA            COMPLETED
```

5.3.1 IM 列式存储的优先级

优先级设置适用于整个表、分区、子分区，但不适用于列。在对象上设置 INMEMORY 属性意味着此对象是 IM 列式存储中的候选对象，但并不意味着数据库会立即将对象存储到 IM 中。Oracle 数据库按如下方式管理存储优先级。

- 按需：默认情况下，INMEMORY PRIORITY 参数为 NONE。在这种情况下，只有当对象被以全表扫描的方式访问后，Oracle 才会将该对象存储到 IM 中。如果是其他访问方式，则不行。
- 基于优先级：如果 INMEMORY PRIORITY 参数不是 NONE，那么 Oracle 会根据优先级顺序将对象存储到 IM 中，此时全表扫描已经不是必要的触发条件了。
  - 在数据库重启后，会自动将对象存储到 IM 中。
  - INMEMORY 对象将根据存储优先级依次存储到 IM 中。例如，INMEMORY PRIORITY CRITICAL > INMEMORY PRIORITY HIGH > INMEMORY PRIORITY LOW。
  - 等待 ALTER TABLE 或 ALTER MATERIALIZED VIEW 的执行结果，直到对象的更改信息记录在 IM 列式存储中。

如表 5-1 所示是 IM 存储优先级的具体说明。

表 5-1 IM 存储优先级的具体说明

PRIORITY 选项	说　明
PRIORITY NONE	对象只有被以全表扫描的方式访问后，才会被存储到 IM 中
PRIORITY LOW	不管对象是否被访问，都会被存储到 IM 中，但是优先级是最低的，永远在队伍的最后
PRIORITY MEDIUM	同上，但是优先级比 LOW 高
PRIORITY HIGH	同上，但是优先级比 MEDIUM 高
PRIORITY CRITICAL	同上，但是优先级是最高的

表 5-1 中的优先级存储顺序，都在 IM 空间足够的情况下才会发生。如果 IM 空间不够，只能等有空闲空间了才会按照优先级顺序依次存储。

5.3.2 IM 列式存储的压缩方法

压缩就是用性能换取空间，但是节省了空间，就会降低性能（CPU 开销更大）。可以通过 V$IM_SEGMENTS 视图和 V$IM_COLUMN_LEVEL 视图查看对象的压缩方法，也可以通过 ALTER 命令修改压缩方法。如果修改的对象正在 IM 中，而修改的是 PRIORITY 以外的属性，那么 Oracle 首先会将该对象从 IM 中删除，再根据存储优先级将对象重新存储到 IM 中。

修改语法：ALTER TABLE TABLE_NAME INMEMORY + 压缩方法

查看 IM 中是否有 CNDBA 表，代码如下：

```
SQL> select segment_name, populate_status from v$im_segments where segment_name = 'CNDBA';
SEGMENT_NAME      POPULATE_STATUS
----------------- ---------------
CNDBA             COMPLETED
```

修改 CNDBA 表的压缩方法，代码如下：

```
SQL> alter table cndba.cndba INMEMORY MEMCOMPRESS FOR QUERY HIGH;
Table altered.
```

再次查看 IM 中是否有 CNDBA 表，代码如下：

```
SQL> select segment_name, populate_status from v$im_segments where segment_name = 'CNDBA';
no rows selected
```

表 5-2 是 IM 列式存储的压缩方法的说明。

表 5-2　IM 列式存储的压缩方法的说明

压缩方法	说明
NO MEMCOMPRESS	不压缩
MEMCOMPRESS FOR DML	对 DML 操作来说性能最好，压缩程度最低
MEMCOMPRESS FOR QUERY LOW	对查询来说性能最好，压缩程度比 FOR DML 高，是默认值

续表

压缩方法	说　　明
MEMCOMPRESS FOR QUERY HIGH	对查询来说性能也不错，同时节省空间，压缩程度比 FOR QUERY LOW 高
MEMCOMPRESS FOR CAPACITY LOW	平衡空间和查询性能，相对更节省空间，压缩程度比 FOR QUERY HIGH 高。这种压缩方法在被扫描之前，Oracle 会先进行解压缩。当指定了 MEMCOMPRESS FOR CAPACITY 而没有指定具体的压缩程度时，默认为该压缩方法
MEMCOMPRESS FOR CAPACITY HIGH	最节省空间，但是查询性能也是最差的

5.3.3 Oralce 压缩顾问

通过调用 DBMS_COMPRESSION.GET_COMPRESSION_RATIO 存储过程预估表或索引的压缩率。在 MOS 上有一篇文章是关于如何使用该存储过程的，并提供了一个分装好的脚本：《How to Use DBMS_COMPRESSION.GET_COMPRESSION_RATIO in 12c》（文档 ID 1589879.1）。

依次输入对象所在表空间、对象所属用户、对象名、压缩类型（如表 5-3 所示，输入对应的值）和行数，代码如下：

```
SQL>set serveroutput on
declare
v_blkcnt_cmp pls_integer;
v_blkcnt_uncmp pls_integer;
v_row_cmp pls_integer;
v_row_uncmp pls_integer;
v_cmp_ratio number;
v_comptype_str varchar2（60）;
begin
dbms_compression.get_compression_ratio（
scratchtbsname => upper（'&ScratchTBS'）,
ownname => upper（'&ownername'）,
objname => upper（'&TableName'）,
subobjname => NULL,
comptype => &compression_type_number,
subset_numrows=> &num_rows,
blkcnt_cmp => v_blkcnt_cmp,
blkcnt_uncmp => v_blkcnt_uncmp,
row_cmp => v_row_cmp,
```

```
    row_uncmp => v_row_uncmp,
    cmp_ratio => v_cmp_ratio,
    comptype_str => v_comptype_str ) ;
    dbms_output.put_linc ('.') ;
    dbms_output.put_line ('OUTPUT: ') ;
    dbms_output.put_line ('Estimated Compression Ratio: '||to_char
(v_cmp_ratio)) ;
    dbms_output.put_line ('Blocks used by compressed sample:
'||to_char (v_blkcnt_cmp)) ;
    dbms_output.put_line ('Blocks used by uncompressed sample:
'||to_char (v_blkcnt_uncmp)) ;
    dbms_output.put_line ('Rows in a block in compressed sample:
'||to_char (v_row_cmp)) ;
    dbms_output.put_line ('Rows in a block in uncompressed sample:
'||to_char (v_row_uncmp)) ;
    end;
    /
```

结果如下：

```
Enter value for scratchtbs: CNDBA
old  10: scratchtbsname => upper ('&ScratchTBS') ,
new  10: scratchtbsname => upper ('CNDBA') ,
Enter value for ownername: lei
old  11: ownname => upper ('&ownername') ,
new  11: ownname => upper ('lei') ,
Enter value for tablename: test
old  12: objname => upper ('&TableName') ,
new  12: objname => upper ('test') ,
Enter value for compression_type_number: 2
old  14: comptype => &compression_type_number,
new  14: comptype => 2,
Enter value for num_rows: 1700
old  15: subset_numrows=> &num_rows,
new  15: subset_numrows=> 1700,
.
OUTPUT:
Estimated Compression Ratio: 9.6
Blocks used by compressed sample: 8
Blocks used by uncompressed sample: 77
Rows in a block in compressed sample: 213
Rows in a block in uncompressed sample: 22
```

脚本中涉及的压缩类型及说明如表 7-3 所示。

表 5-3 脚本中涉及的压缩类型及说明

常 量	类 型	值	说 明
COMP_NOCOMPRESS	NUMBER	1	不压缩
COMP_ADVANCED	NUMBER	2	高级压缩
COMP_QUERY_HIGH	NUMBER	4	对查询的高级压缩
COMP_QUERY_LOW	NUMBER	8	对查询的低级压缩
COMP_ARCHIVE_HIGH	NUMBER	16	对归档操作的高级压缩
COMP_ARCHIVE_LOW	NUMBER	32	对归档操作的低级压缩
COMP_BLOCK	NUMBER	64	压缩的行
COMP_LOB_HIGH	NUMBER	128	对 LOB 操作的高级压缩
COMP_LOB_MEDIUM	NUMBER	256	对 LOB 操作的中级压缩
COMP_LOB_LOW	NUMBER	512	对 LOB 操作的低级压缩
COMP_INDEX_ADVANCED_HIGH	NUMBER	1024	对索引的高级压缩
COMP_INDEX_ADVANCED_LOW	NUMBER	2048	对索引的低级压缩
COMP_RATIO_LOB_MINROWS	NUMBER	1000	估计 LOB 压缩比的对象中所需的最小 LOB 数量
COMP_BASIC	NUMBER	4096	基本压缩级别
COMP_RATIO_LOB_MAXROWS	NUMBER	5000	用于计算 LOB 压缩比的最大 LOB 数
COMP_INMEMORY_NOCOMPRESS	NUMBER	8192	没有使用压缩的 In-Memory
COMP_INMEMORY_DML	NUMBER	16384	对 DML 操作的 In-Memory 压缩
COMP_INMEMORY_QUERY_LOW	NUMBER	32768	对查询的 In-Memory 低级压缩
COMP_INMEMORY_QUERY_HIGH	NUMBER	65536	对查询的 In-Memory 高级压缩
COMP_INMEMORY_CAPACITY_LOW	NUMBER	32768	对优化容量的 In-Memory 低级压缩
COMP_INMEMORY_CAPACITY_HIGH	NUMBER	65536	对优化容量的 In-Memory 高级压缩
COMP_RATIO_MINROWS	NUMBER	1000000	预估 HCC 压缩比率的对象中所需的最小行数
COMP_RATIO_ALLROWS	NUMBER	–1	使用对象中的所有行来估计 HCC 比率

续表

常　量	类　型	值	说　明
OBJTYPE_TABLE	PLS_INTEGER	1	标识压缩比被估计为类型表的对象
OBJTYPE_INDEX	PLS_INTEGER	2	标识压缩比被估计为类型索引的对象

5.3.4　后台进程填充 IMCU

在填充期间，数据库以磁盘的行格式从磁盘读取数据，将行转换为列，然后将数据压缩到 IMCU（In-Memory Compression Units）。

工作进程 Wnnn 负责填充 IM 列式存储中的数据。每个工作进程对来自对象的数据库块的子集进行操作。填充是一种流式机制，可同时压缩数据并将其转换为列式格式。

INMEMORY_MAX_POPULATE_SERVERS 初始化参数指定用于 IM 列式存储区填充的 Wnnn 的最大进程数量。默认情况下，该设置是 CPU_COUNT 值的一半。用户根据自己的环境设置一个合适的值。更多的 Wnnn 进程会有更快的存储速度，但消耗更多的 CPU 资源；相反，更少的 Wnnn 进程会导致填充速度变慢，但是可以减小 CPU 开销。

如果 INMEMORY_MAX_POPULATE_SERVER 参数的值为 0，则表示不启用 IM 填充。

5.3.5　具体操作示例

1. 新表启用 IM 列式存储

在建表时启用 IM 并采用查询的高级压缩方式，代码如下：

```
SQL> create table cndba.new_cndba (id int) inmemory memcompress for query high;
Table created.
```

查看是否启用了 IM，代码如下：

```
SQL> SET LINESIZE 200
SQL> COL TABLE_NAME FORMAT A20
SQL> COL OWNER FORMAT A20
```

```
SQL>select
table_name,owner,inmemory,inmemory_priority,inmemory_compression
from dba_tables where table_name='NEW_CNDBA';

TABLE_NAME OWNER   INMEMORY    INMEMORY_PRIORIT
INMEMORY_COMPRESSION
---------- ---------------- ----------------------------------
NEW_CNDBA  CNDBA      ENABLED       NONE        FOR QUERY HIGH
```

可以看到已经启用了 IM，采用的是 FOR QUERY HIGH 压缩方式。

创建分区表，对每个分区启用 IM，根据不同分区的数据查询和 DML 频率，采用不同的压缩方法，代码如下：

```
SQL>create table cndba.range_sales
   ( prod_id       number (6)
   , cust_id       number
   , time_id       date
   , channel_id     char (1)
   , promo_id      number (6)
   , quantity_sold  number (3)
   , amount_sold    number (10,2)
   )
partition by range (time_id)
  (partition sales_q4_1999
    values less than (to_date('01-jan-2018','dd-mon-yyyy'))
    inmemory memcompress for dml,
  partition sales_q1_2000
    values less than (to_date('01-apr-2018','dd-mon-yyyy'))
    inmemory memcompress for query,
  partition sales_q2_2000
    values less than (to_date('01-jul-2018','dd-mon-yyyy'))
    inmemory memcompress for capacity,
  partition sales_q3_2000
    values less than (to_date('01-oct-2018','dd-mon-yyyy'))
    no inmemory,
  partition sales_q4_2000
    values less than (maxvalue)) ;
```

2. 已存在的表启用 IM 列式存储

在启用了 IM 列式存储的情况下，可以直接修改表的属性来启用 IM。

创建测试表，代码如下：

```
SQL> create table cndba.im as select * from dba_objects;
Table created.
```

查看执行计划，全表扫描，代码如下：

```
SQL>select object_name,object_type from im where object_id>9;
```

执行计划的代码如下：

```
SQL_ID  1c92p1h150px0, child number 0
-------------------------------------
select object_name,object_type from im where object_id>9

Plan hash value: 3775499121

---------------------------------------------------------------
| Id  | Operation        | Name | E-Rows |E-Bytes| Cost (%CPU) |
E-Time|
---------------------------------------------------------------
|   0 | SELECT STATEMENT  |      |        |       |  397 (100)|       |
|*  1 |  TABLE ACCESS FULL| IM   |  72918 |  3560K|   397   (1) |
00:00:01 |
---------------------------------------------------------------

Predicate Information (identified by operation id):
---------------------------------------------------

   1 - filter("OBJECT_ID">9)

Note
-----
   - Warning: basic plan statistics not available. These are only
collected when:
       * hint 'gather_plan_statistics' is used for the statement or
       * parameter 'statistics_level' is set to 'ALL', at session
or system level
```

启用 IM 表的 IM 列式存储，INMEMORY PRIORITY 使用默认设置 NONE，只有用全表扫描方式才能将数据存储到 IM 中。

```
[oracle@18cDG1 ~]$ sqlplus cndba/cndba@dave
SQL*Plus: Release 18.0.0.0.0 - Production on Thu Oct 25 16:00:58
2018
Version 18.3.0.0.0
Copyright (c) 1982, 2018, Oracle.  All rights reserved.
```

```
    Connected to:
    Oracle Database 18c Enterprise Edition Release 18.0.0.0.0 -
Production
    Version 18.3.0.0.0
    SQL> alter table im inmemory;
    Table altered.
    SQL> col table_name for a10
    SQL>col owner for a10
    SQL>select table_name,owner,INMEMORY,INMEMORY_PRIORITY from
dba_tables where table_name='IM';
    TABLE_NAME OWNER     INMEMORY      INMEMORY_PRIORIT
    ---------- ---------- ---------------- ----------------
    IM     CNDBA     ENABLED        NONE
```

查看 IM 中是否有 IM 表，在全表扫描之前，可以看到 IM 表没有被存储到 IM 中，代码如下：

```
    SQL> select segment_name,populate_status from v$im_segments where
segment_name = 'IM';
    no rows selected
    SQL> select count（1） from im;
      COUNT（1）
    ----------
         72926
    SQL> SELECT SEGMENT_NAME, POPULATE_STATUS FROM V$IM_SEGMENTS WHERE
SEGMENT_NAME = 'IM';
    SEGMENT_NAME          POPULATE_STATUS
    --------------------  --------------------
    IM                    COMPLETED
```

再次查看执行计划，全表扫描，代码如下：

```
    SQL>select object_name,object_type from im where object_id>9;
    SQL_ID 1c92p1h150px0, child number 0
    -------------------------------------
    select object_name,object_type from im where object_id>9

    Plan hash value: 3775499121

    ----------------------------------------------------------------
    |Id  |Operation              |Name|E-Rows|E-Bytes|Cost （%CPU）
| E-Time  |
    ----------------------------------------------------------------
    |   0|SELECT STATEMENT        |    |      |     |   16 （100）
```

```
|        |
    |* 1 | TABLE ACCESS INMEMORY FULL| IM   | 72918 | 3560K|   16
(7) | 00:00:01 |
    --------------------------------------------------------------

    Predicate Information (identified by operation id):
    ---------------------------------------------------

      1 - inmemory("OBJECT_ID">9)   #注意这里使用了 IM
          filter("OBJECT_ID">9)

    Note
    -----
      - Warning: basic plan statistics not available. These are only
collected when:
          * hint 'gather_plan_statistics' is used for the statement or
          * parameter 'statistics_level' is set to 'ALL', at session
or system level
```

启用查询的高级压缩，优先级为 HIGH，代码如下：

```
SQL>alter table im inmemory memcompress for query high priority
high;
Table altered.
```

可以通过 DBA_FEATURE_USAGE_STATISTICS 视图查看 IM 列式存储抽取数据的情况，代码如下：

```
SQL>COL NAME FORMAT a25
SQL>SELECT ul.NAME, ul.DETECTED_USAGES
FROM  DBA_FEATURE_USAGE_STATISTICS ul
WHERE ul.VERSION= (SELECT MAX(u2.VERSION)
                  FROM  DBA_FEATURE_USAGE_STATISTICS u2
                  WHERE u2.NAME = ul.NAME
                  AND   ul.NAME LIKE '%Column Store%');
```

3. 已存在的表禁用 IM 列式存储

查看表的 IM 信息，代码如下：

```
SQL> SET LINESIZE 200
SQL> COL TABLE_NAME FORMAT A20
SQL> COL OWNER FORMAT A20
SQL> SELECT
TABLE_NAME,OWNER,INMEMORY,INMEMORY_PRIORITY,INMEMORY_COMPRESSION
```

```
FROM DBA_TABLES WHERE TABLE_NAME='IM';

    TABLE_NAME OWNER     INMEMORY      INMEMORY_PRIORIT
INMEMORY_COMPRESSION
    ---------- ---------------- ----------------------------------
    IM  CNDBA          ENABLED       HIGH          FOR QUERY HIGH
```

禁用 IM 列式存储，代码如下：

```
SQL> alter table im no inmemory;
Table altered.
```

再次查看 IM 信息，代码如下：

```
    SQL>SELECT
TABLE_NAME,OWNER,INMEMORY,INMEMORY_PRIORITY,INMEMORY_COMPRESSION
FROM DBA_TABLES WHERE TABLE_NAME='IM';
    TABLE_NAME OWNER     INMEMORY      INMEMORY_PRIORIT
INMEMORY_COMPRESSION
    ---------- ------ ---------- ----------------------------------
    IM  CNDBA          DISABLED
```

4．外部表启用 IM 列式存储

从 Oracle 18c 开始，支持外部表的 IM 列式存储，操作步骤如下。

（1）从 DBA_TABLES 生成一个 CSV 文件。

创建/tmp/data、/tmp/log、/tmp/bad 三个目录，CSV 文件保存在/tmp/data 目录下，代码如下：

```
[oracle@18cDG1 tmp]$ mkdir /tmp/data
[oracle@18cDG1 tmp]$ mkdir /tmp/log
[oracle@18cDG1 tmp]$ mkdir /tmp/bad
```

生成 CSV 文件，代码如下：

```
    SQL> show con_name
    DAVE
    SQL> SET HEAD OFF
    SQL> SET PAGES 0
    SQL> SET FEEDBACK OFF
    SQL> SET TERMOUT OFF
    SQL> SPOOL /tmp/data/sh_sales.csv
    SQL> SELECT OWNER || ',' || TABLE_NAME || ',' || TABLESPACE_NAME
FROM  dba_tables;
    SQL> SPOOL OFF
```

注意，这里的 SPOOL 文件中仍然包含 SQL 语句，在执行完毕后，手工删除 SPOOL 文件头尾的 SQL 语句即可。

（2）使用 sh_sales.csv 文件，通过以下脚本创建外部表。

用 sys 账户创建 DIRECTORY，并赋权限给 cndba 用户，代码如下：

```
    [oracle@18cDG1 ~]$ sqlplus sys/oracle@dave as sysdba
    SQL*Plus: Release 18.0.0.0.0 - Production on Thu Oct 25 16:54:17
2018
    Version 18.3.0.0.0
    Copyright (c) 1982, 2018, Oracle.  All rights reserved.
    Connected to:
    Oracle Database 18c Enterprise Edition Release 18.0.0.0.0 -
Production
    Version 18.3.0.0.0
    SQL>CREATE OR REPLACE DIRECTORY admin_dat_dir AS '/tmp/data';
    SQL>CREATE OR REPLACE DIRECTORY admin_log_dir AS '/tmp/log';
    SQL>CREATE OR REPLACE DIRECTORY admin_bad_dir AS '/tmp/bad';
    SQL>GRANT READ ON DIRECTORY admin_dat_dir TO cndba;
    SQL>GRANT WRITE ON DIRECTORY admin_log_dir TO cndba;
    SQL>GRANT WRITE ON DIRECTORY admin_bad_dir TO cndba;
```

用 cndba 账户进行连接，并创建外部表，代码如下：

```
    SQL> conn cndba/cndba@dave
    Connected.
    SQL>CREATE TABLE admin_ext_tables
                   (owner          VARCHAR2 (128) ,
                   table_name      VARCHAR2 (128) ,
                   tablespace_name  VARCHAR2 (200)
                  )
        ORGANIZATION EXTERNAL
         (
          TYPE ORACLE_LOADER
          DEFAULT DIRECTORY admin_dat_dir
          ACCESS PARAMETERS
           (
            records delimited by newline
            badfile admin_bad_dir:'empxt%a_%p.bad'
            logfile admin_log_dir:'empxt%a_%p.log'
            fields terminated by ','
            missing field values are null
             ( owner, table_name,tablespace_name
            )
```

```
    )
    LOCATION ('sh_sales.csv')
   )
   REJECT LIMIT UNLIMITED
   INMEMORY;
```

查询 ALL_EXTERNAL_TABLES 视图，验证是否启用了 IM，代码如下：

```
SQL>COL OWNER FORMAT A10
SQL>COL TABLE_NAME FORMAT A15
SQL>SELECT OWNER, TABLE_NAME,
    INMEMORY, INMEMORY_COMPRESSION
FROM  ALL_EXTERNAL_TABLES
WHERE  TABLE_NAME = 'ADMIN_EXT_TABLES';
OWNER     TABLE_NAME     INMEMORY INMEMORY_COMPRESS
---------- --------------- -------- -----------------
CNDBA    ADMIN_EXT_TABLES   ENABLED  FOR QUERY LOW
```

5. 将外部表数据存储到 IM 列式存储

前面只是在外部表中启用 IM，并没有把数据存储到 IM 列式存储。想要查询 IM 中的外部表，必须把 QUERY_REWRITE_INTEGRITY 参数设置为 stale_tolerated。

（1）设置参数并填充数据，代码如下：

```
SQL> alter session set query_rewrite_integrity=stale_tolerated;
Session altered.
SQL> show parameter query_rewrite_integrity
NAME                 TYPE        VALUE
------------------------------------ ------------------------------
query_rewrite_integrity     string          STALE_TOLERATED
SQL> EXEC DBMS_INMEMORY.POPULATE ('CNDBA', 'ADMIN_EXT_TABLES');
PL/SQL procedure successfully completed
```

（2）查看外部表数据是否被存储到 IM 中，代码如下：

```
SQL>COL OWNER FORMAT a10
SQL>COL SEGMENT_NAME FORMAT a20
SQL>COL POPULATE_STATUS FORMAT a15
SQL>SELECT OWNER,SEGMENT_NAME, POPULATE_STATUS
FROM  V$IM_SEGMENTS
WHERE  SEGMENT_NAME = 'ADMIN_EXT_TABLES';
OWNER      SEGMENT_NAME   POPULATE_STATUS
------------------------------ --------------------------
```

```
CNDBA          ADMIN_EXT_TABLES     COMPLETED
```

6. 可刷新外部表

外部表和内部表有所不同，外部表中的数据如果有变动，就需要手动执行命令将外部表中的数据刷新到 IM 列式存储中，操作步骤如下。

（1）向 CSV 文件中插入新数据，代码如下：

```
[oracle@18cDG1 data]$ echo "Dave,CNDBA,CNDBATBS" >>
/tmp/data/sh_sales.csv
```

（2）使用 DBMS_INMEMORY.REPOPULATE 存储过程刷新外部表数据到 IM 中，代码如下：

```
SQL> EXEC DBMS_INMEMORY.REPOPULATE('CNDBA', 'ADMIN_EXT_TABLES');
PL/SQL procedure successfully completed.
```

在刷新数据的过程中，所有对该表的查询都不会成功。

7. 对虚拟列启用、禁用 IM

可以为内部表中的各个列指定 INMEMORY 子句。外部表不支持在列级别指定 INMEMORY 子句。对于内部表来说，IM 虚拟列和非虚拟列都支持 IM 存储。

（1）IM 虚拟列

IM 虚拟列与其他列相似，只不过它的值是通过表达式计算得来的。在 IM 列式存储中存储预先计算的 IM 虚拟列值可以提高查询性能。表达式中可以包括同一个表的列、常量、SQL 函数和用户定义的 PL / SQL 函数（仅限于 DETERMINISTIC）。虚拟列无法直接修改和插入数据。虚拟列或 IM 表达式最多可以对每个 IM 对象中的 1000 列进行计算。

可以设置 INMEMORY_VIRTUAL_COLUMNS 参数来控制 IM 虚拟列是否自动存储到 IM 中，参数取值如下。

- MANUAL：是默认值，如果一个表启用了 IM 存储，那么该表上的虚拟列不会自动存储到 IM 中，除非指定 INMEMORY 子句。
- ENABLE: 如果一个表启用了 IM 存储，那么该表上的虚拟列将自动存储到 IM 中，除非指定 NO INMEMORY 子句。可以为虚拟列单独指定压缩方式，该压缩方式不同于表的压缩方式。

（2）启用 IM 虚拟列

将参数 INMEMORY_VIRTUAL_COLUMNS 设置为 ENABLE，代码如下：

```
SQL> show parameter inmemory_virtual_columns
NAME                    TYPE        VALUE
------------------------------ ------------------------------
inmemory_virtual_columns    string          MANUAL
SQL> alter system set inmemory_virtual_columns=enable scope=both;
System altered.
```

为 IM 中的表 cndba 新增一个虚拟列，代码如下：

```
SQL> alter table cndba add(add_blocks as (round(object_id)+1000));
Table altered.
```

查看 IM 虚拟列是否被存储到 IM 中，代码如下：

```
SQL> SET LINESIZE 200
SQL> COL TABLE_NAME FORMAT a20
SQL> COL COLUMN_NAME FORMAT a20
SQL> SELECT TABLE_NAME, COLUMN_NAME, INMEMORY_COMPRESSION FROM
V$IM_COLUMN_LEVEL WHERE TABLE_NAME = 'CNDBA' AND
COLUMN_NAME='ADD_BLOCKS' ORDER BY COLUMN_NAME;
TABLE_NAME      COLUMN_NAME   INMEMORY_COMPRESSION
-------------------------------------------------------------------
CNDBA           ADD_BLOCKS    UNSPECIFIED
```

代码中的 UNSPECIFIED 表示没有单独指定压缩方式，即压缩方式和所在表的压缩方式一样。

（3）修改 IM 虚拟列的压缩方式

继续上面的例子，修改 ADD_BLOCKS 虚拟列的压缩方式，代码如下：

```
SQL> alter table cndba inmemory memcompress for query high
(add_blocks);
Table altered.
```

查看压缩方式是否发生了变化，代码如下：

```
SQL> SELECT TABLE_NAME, COLUMN_NAME, INMEMORY_COMPRESSION FROM
V$IM_COLUMN_LEVEL WHERE TABLE_NAME = 'CNDBA' AND
COLUMN_NAME='ADD_BLOCKS' ORDER BY COLUMN_NAME;
TABLE_NAME      COLUMN_NAME   INMEMORY_COMPRESSION
-------------------------------------------------------------------
CNDBA           ADD_BLOCKS    FOR QUERY HIGH
```

8．对表中部分列启用 IM 列式存储

从 Oracle 12.2 开始，可以对没有启用 IM 列式存储的对象中的列启用 IM 列式存储，即只对表中的部分列（查询频繁的列）启用 IM 列式存储。

创建一个分区表 T，且不启用 IM 列式存储，代码如下：

```
SQL>CREATE TABLE t （c1 NUMBER, c2 NUMBER, c3 NUMBER） NO INMEMORY
  PARTITION BY LIST （c1）
     ( PARTITION p1 VALUES （0）,
      PARTITION p2 VALUES （1）,
      PARTITION p3 VALUES （2） );
Table created.
```

查询分区表 T 中的列是否启用了 IM 列式存储，代码如下：

```
SQL> SELECT TABLE_NAME, COLUMN_NAME, INMEMORY_COMPRESSION FROM
V$IM_COLUMN_LEVEL WHERE  TABLE_NAME = 'T' ORDER BY COLUMN_NAME;
no rows selected
```

对 c3 列执行 NO INMEMORY 子句，控制 c3 列不启用 IM 列式存储，代码如下：

```
SQL> ALTER TABLE t NO INMEMORY （c3）;
Table altered.
```

注意，这里在禁用 c3 列的同时，会启用其他列。所以，如果只想启用某列，就需要把其他列禁用掉。

再次查看列的 IM 列式存储启用情况，代码如下：

```
SQL> SELECT TABLE_NAME, COLUMN_NAME, INMEMORY_COMPRESSION FROM
V$IM_COLUMN_LEVEL WHERE  TABLE_NAME = 'T' and owner='CNDBA' ORDER BY
COLUMN_NAME;
TABLE_NAME       COLUMN_NAME    INMEMORY_COMPRESSION
---------------- -------------------------------------
T              C1         DEFAULT
T              C2         DEFAULT
T              C3         NO INMEMORY
```

对 p3 分区启用 IM 列式存储，由于禁用了 c3 列的 IM 列式存储，所以现在 IM 中的 p3 分区中不包含 c3 列，代码如下：

```
SQL> alter table t modify partition p3 inmemory priority critical;
Table altered.
```

对分区表 T 启用 IM 列存储，代码如下：

```
SQL> alter table t inmemory;
Table altered.
```

再次查看列的 IM 列式存储的启用情况，可以看到 c3 列还是没有启用 IM 列式存储，c1、c2 列使用默认压缩方式，代码如下：

```
SQL> SELECT TABLE_NAME, COLUMN_NAME, INMEMORY_COMPRESSION FROM V$IM_COLUMN_LEVEL WHERE  TABLE_NAME = 'T' and owner='CNDBA' ORDER BY COLUMN_NAME;
TABLE_NAME       COLUMN_NAME    INMEMORY_COMPRESSION
-------------------------------------------------------------
T            C1         DEFAULT
T            C2         DEFAULT
T            C3         NO INMEMORY
```

为 c1、c2 列指定不同的压缩方式，代码如下：

```
SQL> alter table t inmemory memcompress for capacity high （c1） inmemory memcompress for capacity low （c2）;
Table altered.
```

再次查看列的 IM 列式存储的启用情况，可以看到 c1、c2 列使用了不同的压缩方式，代码如下：

```
SQL> SELECT TABLE_NAME, COLUMN_NAME, INMEMORY_COMPRESSION FROM V$IM_COLUMN_LEVEL WHERE  TABLE_NAME = 'T' and owner='CNDBA' ORDER BY COLUMN_NAME;
TABLE_NAME       COLUMN_NAME    INMEMORY_COMPRESSION
-------------------------------------------------------------------
T            C1         FOR CAPACITY HIGH
T            C2         FOR CAPACITY LOW
T            C3         NO INMEMORY
```

如果现在将整个分区表 T 禁用 IM 列式存储，那么涉及该表的所有列都会从 IM 列式存储中被删除，代码如下：

```
SQL> alter table t no inmemory;
Table altered.
SQL> SELECT TABLE_NAME, COLUMN_NAME, INMEMORY_COMPRESSION FROM V$IM_COLUMN_LEVEL WHERE  TABLE_NAME = 'T' and owner='CNDBA' ORDER BY COLUMN_NAME;
no rows selected
```

9. 启用、禁用表空间的 IM 列式存储

表空间级 IM 的语法和表级 IM 的语法没有区别，只要把 ALTER TABLE、CREATE TABLE {INMEMORY|NO INMEMORY}替换为 ALTER TABLESPACE、CREATE TABLESPACE DEFAULT {INMEMORY|NO INMEMORY}，其他参数保持不变。

默认情况下，如果表空间启用了 IM 列式存储，那么该表空间中的所有表、物化视图都将启用 IM 列式存储。表和物化视图可以单独设置其他属性，如压缩方式。

（1）启用 IM 列式存储

创建一个新的表空间并启用 IM 列式存储，代码如下：

```
    SQL> create tablespace data datafile '/u01/app/oracle/oradata/
CNDBA/dave/data01.dbf' size 40m online default inmemory;
    Tablespace created.
```

也可以对一个已存在的表空间启用 IM 列式存储，代码如下：

```
    SQL> alter tablespace users default inmemory memcompress for
capacity high priority low;
    Tablespace altered.
```

在该表空间上创建一个对象，看看是否会启用 IM 列式存储，代码如下：

```
    SQL> create table data_table (id int) tablespace data;
    Table created.
    SQL> select
table_name,owner,inmemory,inmemory_priority,inmemory_compression
from dba_tables where table_name='DATA_TABLE';
    TABLE_NAME       OWNER    INMEMORY INMEMORY_PRIORIT
INMEMORY_COMPRESSION
    --------------- ------ -------------- -------------------------
    DATA_TABLE       LEI  ENABLED   NONE           FOR QUERY LOW
```

可以看到，表空间中的对象会自动启用 IM 列式存储，并且会继承表空间的 IM 列式存储的相关属性。

（2）禁用 IM 列式存储

禁用表空间的 IM 列式存储，会将该表空间中的所有对象从 IM 列式存储中删除，代码如下：

```
SQL> alter tablespace data default no inmemory ;
Tablespace altered.
```

10. 启用、禁用物化视图的 IM 列式存储

物化视图的 IM 语法和表、表空间的 IM 语法相似，其语法如下：

```
CREATE MATERIALIZED VIEW  {INMEMORY|NO INMEMORY}
```

或

```
ALTER MATERIALIZED VIEW  {INMEMORY|NO INMEMORY}
```

（1）启用 IM 列式存储

创建新的物化视图并启用 IM 列式存储，代码如下：

```
SQL> create materialized view mat inmemory as select * from cndba;
Materialized view created.
```

对已存在的物化视图启用 IM 列式存储，代码如下：

```
SQL> alter materialized view mat inmemory priority high;
Materialized view altered.
SQL> col segment_name for a10
SQL> select owner, segment_name, inmemory_priority,
inmemory_compression from v$im_segments;
OWNER      SEGMENT_NA INMEMORY_PRIORIT INMEMORY_COMPRESSION
---------- ---------------- -----------------------------------
CNDBA      MAT        HIGH             FOR CAPACITY HIGH
CNDBA      CNDBA      LOW         FOR QUERY HIGH
```

（2）禁用 IM 列式存储

禁用 IM 列式存储的代码如下：

```
SQL> alter materialized view mat no inmemory;
Materialized view altered.
SQL> SELECT OWNER, SEGMENT_NAME, INMEMORY_PRIORITY,
INMEMORY_COMPRESSION FROM V$IM_SEGMENTS;
OWNER      SEGMENT_NA INMEMORY_PRIORIT INMEMORY_COMPRESSION
---------- ----------- ----------------------------------
CNDBA      CNDBA      LOW         FOR QUERY HIGH
```

5.3.6 强制存储数据到 IM 中

默认情况下，Oracle 会根据优先级顺序或是否执行了全表扫描来决定是否把对象存储到 IM 中，但是 Oracle 也提供了将对象强制存储到 IM 中的方法，

即通过调用 DBMS_INMEMORY.POPULATE 存储过程来实现。

创建表 USTC 并启用 IM 列式存储，代码如下：

```
    SQL> create table ustc inmemory priority low as select * from
dba_objects;
    Table created.
```

查看 V$IM_SEGMENTS 视图中是否有记录，代码如下：

```
    SQL>col owner format a10;
    SQL>col name format a25;
    SQL>col status format a10;
    SQL>select owner, segment_name name,populate_status status from
v$im_segments where  segment_name = 'USTC';
    no rows selected
```

强制存储到 IM 中，代码如下：

```
    SQL> EXEC DBMS_INMEMORY.POPULATE ('CNDBA', 'USTC') ;
    PL/SQL procedure successfully completed.
```

再次查询，代码如下：

```
    SQL> select owner, segment_name name,populate_status status from
v$im_segments where  segment_name = 'USTC';
    OWNER      NAME                      STATUS
    ---------- ------------------------- ----------
    CNDBA      USTC                      COMPLETED
```

5.4　自动管理 IM 列式存储中的对象

可以使用 Automatic Data Optimization（ADO）和 Automatic In-Memory（AIM）动态管理 IM 列式存储中的对象。ADO 和 AIM 都使用 Heat Map，Heat Map 会跟踪块和段的数据访问频率。ADO 和 Heat Map 是 ILM（Information Lifecycle Management）的一部分，ILM 是一组用于管理从数据创建到数据存档、再到数据删除的数据进程和策略。

5.4.1　IM 列式存储启用 ADO

ADO 会创建策略并根据这些策略自动执行操作，从而实现 ILM。ADO 通过策略来管理 IM 列式存储。只能在段级别使用 INMEMORY 子句来创建 ADO

策略。数据库将 ADO 策略视为对象的属性。ADO 策略是数据库级别的，而不是实例级别的。

Oracle 支持以下类型用于 IM 的 ADO 策略。

- INMEMORY 策略：启用对象 IM。
- 修改压缩方式策略：为 IM 中的对象修改压缩方式。
- NO INMEMORY 策略：禁用对象 IM 列式存储。

Oracle 可以通过以下方式来决定执行策略的时间。

- 自对象被修改后的指定天数：该天数取自 DBA_HEAT_MAP_SEGMENT 视图中的 SEGMENT_WRITE_TIME 列。
- 自对象被访问后的指定天数：该天数是 DBA_HEAT_MAP_SEGMENT 视图中 SEGMENT_WRITE_TIME、FULL_SCAN 和 LOOKUP_SCAN 列中最大的一个值。
- 自对象被创建后的指定天数：从 DBA_OBJECTS 的 CREATED 列中获取天数。
- 用户定义的函数返回一个布尔值。

5.4.2 ADO 和 IM 结合的优点

可以创建合适的策略，当对象访问较少时就将对象从 IM 列式存储中逐出，访问频繁时再将对象存储到 IM 列式存储中，从而提高查询性能。ADO 使用 Heat Map 记录的信息来管理 IM 列式存储。

1. INMEMORY 策略

在现在的大部分数据库中，对象被创建后都会进行大量的数据插入。而在此期间，为了不影响数据库性能，通常不会对该对象进行其他操作。此时可以通过 ADO，待数据插入完成后再将该对象存储到 IM 列式存储中。例如，每天向表中添加分区，那么就可以创建一个策略，在创建分区一天后再将分区存储到 IM 列式存储中，代码如下：

```
SQL>ALTER TABLE sales MODIFY PARTITION cndba_2018_week10 ILM ADD
POLICY SET INMEMORY MEMCOMPRESS FOR QUERY
    PRIORITY HIGH  AFTER 1 DAYS OF CREATION;
```

同样，也可以在对象创建 60 天后再将其存储到 IM 列式存储中，代码如下：

```
SQL>ALTER TABLE cndba_test ILM ADD POLICY SET INMEMORY MEMCOMPRESS
FOR QUERY PRIORITY CRITICAL AFTER 60 DAYS OF CREATION;
```

2．修改压缩策略

如果想根据访问频率来压缩数据，则可以考虑使用修改压缩策略。例如，当一个对象距离最后一次 DML 操作时间超过两天时，就可以对该对象执行压缩操作，从而节省内存空间，代码如下：

```
SQL>ALTER TABLE EMP ILM ADD POLICY MODIFY INMEMORY MEMCOMPRESS FOR
QUERY HIGH AFTER 2 DAYS OF NO MODIFICATION;
```

如果该对象没有被存储到 IM 列式存储中，则压缩策略仅更改压缩属性；如果该对象已被存储到 IM 列式存储中，则 ADO 使用新的压缩级别重新将该对象存储到 IM 中；如果该对象还没有启用 IM 列式存储，则数据库将忽略该策略。

3．NO INMEMORY 策略

一个对象（分区）经过一段时间后访问量就会减少，可以指定 NO INMEMORY 策略将 IM 中访问量很少的对象从 IM 中删除，从而将 IM 空间留给其他更需要该空间的对象。

例如，一个表的分区如果在 7 天之内都没有被访问，那么就将该分区禁用 IM 列式存储，代码如下：

```
SQL>ALTER TABLE sales MODIFY PARTITION cndba_2015_q1 ILM ADD POLICY
NO INMEMORY AFTER 7 DAYS OF NO ACCESS;
```

4．ADO、IM 相关的初始化参数

ADO、IM 相关的初始化参数及其说明如表 5-4 所示。

表 5-4　ADO、IM 相关的初始化参数及其说明

初始化参数	说　明
COMPATIBLE	要使用 ADO 来管理 IM 列式存储，该参数值必须是 12.2.0 及以上
HEAT_MAP	要启用 Heat Map 和 ADO 特性，该参数值要设置为 ON
INMEMORY_SIZE	IM 内存大小。要启用 IM，该参数值必须是非 0

5．ADO、IM 相关的视图

ADO、IM 相关的视图及其说明如表 5-5 所示。

表 5-5 ADO、IM 相关的视图及其说明

视　图	说　明
DBA_HEAT_MAP_SEG_HISTOGRAM	显示用户可见的所有段的访问信息
DBA_HEAT_MAP_SEGMENT	显示用户可见的所有段的最新的访问时间
DBA_HEATMAP_TOP_OBJECTS	默认显示前 10 000 个对象的热图信息
DBA_HEATMAP_TOP_TABLESPACES	显示前 10 000 个表空间的热图信息
DBA_ILMDATAMOVEMENTPOLICIES	显示特定于数据库中 ADO 策略的数据移动相关属性的信息。action_type 列描述与 IM 列式存储相关的策略。可能的值是 COMPRESSION，STORAGE、EVICT 和 ANNOTATE
V$HEAT_MAP_SEGMENT	显示段的实时访问信息

5.4.3 配置自动 IM

自动 IM 通过使用访问跟踪、列统计信息和其他相关统计信息来管理 IM 列式存储中的对象。如果 IM 列式存储空间已满，并且 IM 中有其他频繁访问的对象，那么 IM 列式存储就会删除不活动的对象，留出空间给其他对象。但是，如果 IM 列式存储被设置成必须保存所有 INMEMORY 段，则自动 IM 不会执行任何操作。

从 Oracle 18c 开始，自动 IM 透明地自动管理 IM 列式存储，通过删除访问很少的段来确保始终存储工作数据集（Working Data Set）。此功能有以下优点。

- 提升性能：通过删除访问很少的段来减轻内存压力，因为工作数据集驻留在 IM 列式存储中，所以自动 IM 提高了工作负载的性能。
- 易于管理：自动 IM 减少了用户的干预。

1. IM 运行机制

只有优先级被设置为 NONE 的对象，才会被自动删除。IM 运行机制基本的处理过程如下。

（1）如果对象存储到 IM 中失败，意味着 IM 内存空间已经用完。

（2）数据库通过已存储到 IM 中的段的 Heat Map 统计信息来找出要删除的对象。

（3）对于集合中的每个段，数据库会检查是否为各个段单独启用了 ADO 策略：

- 如果启用的策略要求该段保持填充状态，则 ADO 策略将覆盖自动 IM；
- 如果没有制定策略来保留段，那么自动 IM 就会执行任务来删除段。

（4）Wnnn 进程删除通过上述检查的任何对象，释放 IM 列式存储中的空间。被删除的对象仍然保留 INMEMORY 的相关属性。

2. INMEMORY_AUTOMATIC_LEVEL 参数说明

要启用自动 IM，需要设置 INMEMORY_AUTOMATIC_LEVEL 初始化参数，该参数有以下三个选项。

- OFF（默认）：禁用自动内存。
- LOW：在内存压力下，数据库将从 IM 列式存储中移除不常用的对象。
- MEDIUM：该选项有一个额外的优化，可将因内存压力而没有被存储到 IM 中的对象优先存储到 IM 中。

Oracle 建议为工作数据集配置足够的内存以满足 IM 列式存储。Oracle 给了一个参考标准，即 5 KB 乘以 SGA 内存的 INMEMORY 段的数量。例如，如果 IM 列式存储中有 10 000 段，则为自动 IM 预留 50 MB 的共享池。

启用自动 IM 的语法如下：

```
SQL>ALTER SYSTEM SET INMEMORY_AUTOMATIC_LEVEL = 'LOW|MEDIUM'
SCOPE=BOTH;
```

3. DBMS_INMEMORY_ADMIN 说明

使用 DBMS_INMEMORY_ADMIN 包可以指定自动 IM 检查统计信息的时间范围。例如，可以指定自动 IM 仅检查过去一个月或过去一周的统计信息（默认一个月）。

查看当前的配置，代码如下：

```
SQL> VARIABLE b_interval NUMBER
BEGIN
DBMS_INMEMORY_ADMIN.AIM_GET_PARAMETER
(DBMS_INMEMORY_ADMIN.AIM_STATWINDOW_DAYS, :b_interval);
END;
/
PL/SQL procedure successfully completed.
SQL> PRINT b_interval
B_INTERVAL
----------
    31
```

可以通过以下方式修改时间，如修改为 7 天，代码如下：

```
SQL>EXEC DBMS_INMEMORY_ADMIN.AIM_SET_PARAMETER
(DBMS_INMEMORY_ADMIN.AIM_STATWINDOW_DAYS, 7 );
```

第6章

优化 IM 查询

6.1 优化 IM 表达式

表达式由一个或多个值、操作符、SQL 语句或函数组成，可以返回一个值。ESS （Expression Statistics Store）会自动跟踪使用频率高的表达式。可以通过 DBMS_INMEMORY_ADMIN 包来捕获指定时间范围内使用频率前 20 的表达式，并将它们作为隐藏虚拟列存储到 IM 中，也可以从 IM 中删除。Oracle 只会记录已经存储到 IM 中的表（可以是部分列）的表达式。

6.1.1 表达式的捕获间隔

从 Oracle 18c 开始，DBMS_INMEMORY_ADMIN.IME_CAPTURE_EXPRESSIONS 存储过程中的 snapshot 参数可以指定以下值来定义捕获表达式的时间间隔。

- CUMULATIVE：数据库创建以来的统计信息。
- CURRENT：过去 24 小时的统计信息。
- WINDOW：从手动调用 IME_OPEN_CAPTURE_WINDOW 存储过程打开捕获的这个时间开始，到调用 IME_CLOSE_CAPTURE_WINDOW 存储过程结束捕获的这段时间。可以使用查询视图 DBA_EXPRESSION_STATISTICS，条件是 SNAPSHOT = 'WINDOW'。

调用方式如下：

```
SQL>EXEC DBMS_INMEMORY_ADMIN.IME_OPEN_CAPTURE_WINDOW ();
SQL>EXEC DBMS_INMEMORY_ADMIN.IME_CLOSE_CAPTURE_WINDOW ();
```

6.1.2　SYS_IME 虚拟列

在捕获期间，Oracle 将 20 个最流行的表达式作为隐藏的 SYS_IME 虚拟列添加到各自的表中，并启用默认的 INMEMORY 和压缩方式。对于之前使用频率高的表达式，现在使用频率不那么高了，则会从 IM 中删除（NO INMEMORY）它。

无论表是否启用 IM，表的最大 SYS_IME 列数均为 50。当一个表中有 50 个表达式后，数据库就不再添加新的 SYS_IME 列。如果想添加新的表达式，则必须使用 DBMS_INMEMORY.IME_DROP_EXPRESSIONS 存储过程或 DBMS_INMEMORY_ADMIN.IME_DROP_ALL_EXPRESSIONS 存储过程来删除之前的 SYS_IME 虚拟列。

另外需要注意的是，一个表中最多可以有 1000 个列启用 IM，包括虚拟列。如果一个表中已经有 980 个非虚拟列启用了 IM，那么最多只能再添加 20 个虚拟列。

6.1.3　管理 IM 表达式

可以使用 DBMS_INMEMORY_ADMIN 包、DBMS_INMEMORY 包和 INMEMORY_EXPRESSIONS_USAGE 初始化参数来控制 IM 表达式的行为。

1．INMEMORY_EXPRESSIONS_USAGE 参数

INMEMORY_EXPRESSIONS_USAGE 参数用来控制存储哪种类型的 IM 表达式。INMEMORY_VIRTUAL_COLUMNS 参数可控制普通（非隐藏）虚拟列的数量。

INMEMORY_EXPRESSIONS_USAGE 参数有以下几个值。

- ENABLE（默认）：数据库将静态和动态 IM 表达式存储到 IM 列式存储中，但是会增加某些表的内存占用空间。
- STATIC_ONLY：静态配置使 IM 列式存储可以缓存 OSON（二进制 JSON）列，这些列用 IS_JSON 检查约束标记。在内部，OSON 列是名为 SYS_IME_OSON 的隐藏虚拟列。

- DYNAMIC_ONLY：数据库仅存储经常使用的表达式，这些表达式作为 SYS_IME 隐藏虚拟列被添加到表中，但是会增加表的内存占用空间。
- DISABLE：不会存储任何 IM 表达式。

修改 INMEMORY_EXPRESSIONS_USAGE 参数的参数值不会立刻对已经存储的表达式产生影响，只有在下次重新存储表达式时才会生效。修改语法如下：

```
SQL>ALTER SYSTEM SET INMEMORY_EXPRESSIONS_USAGE='DISABLE';
```

2. DBMS_INMEMORY_ADMIN 包和 DBMS_INMEMORY 包

表 6-1 是 DBMS_INMEMORY_ADMIN 包和 DBMS_INMEMORY 包中的存储过程及其说明。

表 6-1　DBMS_INMEMORY_ADMIN 包和 DBMS_INMEMORY 包中的存储过程及其说明

包　名	存储过程名	说　明
DBMS_INMEMORY_ADMIN	IME_OPEN_CAPTURE_WINDOW	开始捕获表达式
DBMS_INMEMORY_ADMIN	IME_CLOSE_CAPTURE_WINDOW	停止捕获表达式
DBMS_INMEMORY_ADMIN	IME_GET_CAPTURE_STATE	返回表达式捕获窗口的当前捕获状态和最近修改的时间戳
DBMS_INMEMORY_ADMIN	IME_CAPTURE_EXPRESSIONS	捕获指定时间范围内数据库中最常访问的 20 个表达式
DBMS_INMEMORY_ADMIN	IME_POPULATE_EXPRESSIONS	强制存储最近一次调用 IME_CAPTURE_EXPRESSIONS 过程中捕获的 IM 表达式
DBMS_INMEMORY_ADMIN	IME_DROP_ALL_EXPRESSIONS	删除所有 SYS_IME 虚拟列
DBMS_INMEMORY	IME_DROP_EXPRESSIONS	删除某个表中指定的一组虚拟列

6.1.4　捕获 IM 表达式

启用捕获 IM 表达式的前提条件如下。

- INMEMORY_EXPRESSIONS_USAGE 参数值不能是 DISABLE。
- 启用 IM 列式存储。
- COMPATIBLE 参数值必须是 12.2.0 及以上。
- 指定捕获表达式的时间窗口。

下面通过具体的示例来说明如何使用捕获IM表达式。

1. 自定义捕获IM表达式并存储IM表达式

执行捕获IM表达式的命令，代码如下：

```
SQL> EXEC DBMS_INMEMORY_ADMIN.IME_OPEN_CAPTURE_WINDOW();
PL/SQL procedure successfully completed.
```

查看状态，代码如下：

```
SQL> VARIABLE b_state VARCHAR2(25)
SQL> VARIABLE b_time VARCHAR2(10)
SQL> EXECUTE DBMS_INMEMORY_ADMIN.IME_GET_CAPTURE_STATE
(:b_state, :b_time)
SQL> PRINT b_state b_time
B_STATE
-------------------------------------------
OPEN
B_TIME
--------------------------------------------
25-OCT-18
```

执行表达式查询，并且多执行几次，代码如下：

```
SQL>select round((OBJECT_ID+100)/12,2),OBJECT_NAME from cndba;
```

关闭捕获表达式，代码如下：

```
SQL>EXEC DBMS_INMEMORY_ADMIN.IME_CLOSE_CAPTURE_WINDOW();
PL/SQL procedure successfully completed.
```

查看是否有捕获表达式的统计信息，代码如下：

```
SQL>COL OWNER FORMAT A6
SQL>COL TABLE_NAME FORMAT A9
SQL>COL COUNT FORMAT 99999
SQL>COL CREATED FORMAT A30
SQL>COL EXPRESSION_TEXT FORMAT A29
SQL>SELECT OWNER, TABLE_NAME, EVALUATION_COUNT AS COUNT, CREATED,
EXPRESSION_TEXT FROM  DBA_EXPRESSION_STATISTICS WHERE  SNAPSHOT =
'WINDOW' AND OWNER='CNDBA';
OWNER  TABLE_NAM  COUNT CREATED              EXPRESSION_TEXT
------ -------------------------------------------------------
CNDBA  CNDBA    72918 25-OCT-18              "OBJECT_NAME"
CNDBA  CNDBA    72918 25-OCT-18              "OBJECT_ID"
```

```
CNDBA  CNDBA   72918 25-OCT-18              ROUND
(("OBJECT_ID"+100) /12,2)
```

使用 IME_CAPTURE_EXPRESSIONS 捕获表达式，代码如下：

```
SQL>EXEC DBMS_INMEMORY_ADMIN.IME_CAPTURE_EXPRESSIONS('WINDOW');
PL/SQL procedure successfully completed.
```

查看 IM 中是否有捕获表达式，代码如下：

```
SQL>COL OWNER FORMAT a6
SQL>COL TABLE_NAME FORMAT a9
SQL>COL COLUMN_NAME FORMAT a25
SQL>SET LONG 50
SQL>SET LINESIZE 150
SQL>SELECT OWNER, TABLE_NAME, COLUMN_NAME, SQL_EXPRESSION FROM
DBA_IM_EXPRESSIONS;
OWNER  TABLE_NAM COLUMN_NAME        SQL_EXPRESSION
-----------------------------------------------------------------
CNDBA  CNDBA  SYS_IME0001000000288F67  ROUND (("OBJECT_ID"+100)
/12,2)
```

可以看到，表达式已经被存储到 IM 中。

如果需要，也可以强制将所有捕获到的表达式存储到 IM 中，代码如下：

```
SQL> EXEC DBMS_INMEMORY_ADMIN.IME_POPULATE_EXPRESSIONS () ;
PL/SQL procedure successfully completed.
```

2．捕获过去 24 小时的前 20 个表达式

捕获过去 24 小时的前 20 个表达式和自定义捕获表达式的唯一区别就是，不用手动打开、关闭捕获表达式的时间窗口。

执行一些表达式查询，代码如下：

```
SQL>select 12* (NVL (OBJECT_ID,1) +100) ,OBJECT_NAME from cndba;
```

捕获过去 24 小时的表达式，代码如下：

```
SQL> EXEC DBMS_INMEMORY_ADMIN.IME_CAPTURE_EXPRESSIONS
('CURRENT') ;
PL/SQL procedure successfully completed.
```

将表达式存储到 IM 中，代码如下：

```
SQL> EXEC DBMS_INMEMORY_ADMIN.IME_POPULATE_EXPRESSIONS () ;
PL/SQL procedure successfully completed.
```

查看 IM 中的表达式，代码如下：

```
SQL>SELECT OWNER, TABLE_NAME, COLUMN_NAME, SQL_EXPRESSION FROM
DBA_IM_EXPRESSIONS;
```

3. 删除 IM 中所有的表达式

可以根据用户名、表名、SYS_IME 列名来删除 IM 中的表达式。

查看当前 IM 中的表达式，代码如下：

```
SQL> select owner, table_name, column_name, sql_expression from
dba_im_expressions;
OWNER  TABLE_NAM COLUMN_NAME          SQL_EXPRESSION
------ --------- --------------------------------------------
CNDBA  CNDBA  SYS_IME0001000000288F67  ROUND(("OBJECT_ID"+100)
/12,2)
```

按 SYS_IME 列名来删除表达式，代码如下：

```
SQL>EXEC DBMS_INMEMORY.IME_DROP_EXPRESSIONS('CNDBA',
'CNDBA','SYS_IME0001000000288F67');
```

按表名来删除表达式，代码如下：

```
EXEC DBMS_INMEMORY.IME_DROP_EXPRESSIONS('CNDBA', 'CNDBA');
```

删除全部表达式，代码如下：

```
DBMS_INMEMORY_ADMIN.IME_DROP_ALL_EXPRESSIONS();
```

6.2　使用连接组优化连接

连接组（Join Group）是用户创建的字典对象，列出了一个或多个连接的列。当启用了 IM 列式存储后，Oracle 会使用连接组来优化 IM 中表之间的连接操作。

连接组是一组经常连接的表中的列，可以包含一列或多列，最多可以有 255 列。连接组可以包含不同表的不同列，但是同一列不能同时在多个连接组中。

例如，如果表 T1 中的列 create_time 和表 T2 中的列 create_time 经常有连接操作，那么就可以创建一个连接组（T1（create_time），T2（create_time））。需要注意的是，Oracle ADG 的备库不支持连接组。

连接组有如下优点。

- Oracle 可以直接对压缩数据进行操作。

- 避免了对连接键进行散列和查看散列表，否则需要比较散列表的行数和散列行的散列键。
- Oracle 会通过内部数据字典来创建连接组，数据字典代码密集且具有固定长度，这样就大大提升了空间利用率。
- 在无法使用 Bloom filter 时，有时可以使用连接组来优化查询。

6.2.1 创建连接组

可以通过 CREATE INMEMORY JOIN GROUP 语句创建连接组，它的元数据存储在数据字典中。Oracle 不会立即编译通用数据字典（Common Dictionary），只有连接组中的表在下次被重新存储到 IM 中时，Oracle 才会编译通用数据字典。

所有对连接组的操作（创建、修改或删除）都会造成其引用的表无效，因此，Oracle 建议在首次把表存储到 IM 之前创建连接组，这样就不会造成表的无效。

下面的示例是 USTC 表和 CNDBA 表以 object_id 作为连接条件进行查询，连接组对于查询性能的提升非常有效。

创建查询连接组，代码如下：

```
    SQL> CREATE INMEMORY JOIN GROUP cndba_jg （ustc（object_id），cndba
（object_id））;
    Join group created.
```

查询连接组的定义代码如下：

```
    SQL> COL JOINGROUP_NAME FORMAT a20
    SQL> COL TABLE_NAME FORMAT a12
    SQL> COL COLUMN_NAME FORMAT a20
    SQL> SELECT JOINGROUP_NAME, TABLE_NAME, COLUMN_NAME, GD_ADDRESS
FROM DBA_JOINGROUPS;
    JOINGROUP_NAME       TABLE_NAME   COLUMN_NAME        GD_ADDRESS
    -------------------- --- -------------------- ----------------
    CNDBA_JG     USTC        OBJECT_ID
    CNDBA_JG     CNDBA       OBJECT_ID
```

将表存储到 IM 中，如果存储之前表已经在 IM 中，那么需要将表重新存储到 IM 中；如果存储之前表没有启用 IM，那么需要将表存储到 IM 中，代码如下：

```
SQL> ALTER TABLE cndba.cndba INMEMORY;
SQL>ALTER TABLE cndba.ustc INMEMORY;
SQL>EXEC DBMS_INMEMORY.POPULATE（'CNDBA', 'CNDBA'）;
SQL>EXEC DBMS_INMEMORY.POPULATE（'CNDBA', 'USTC'）;
```

6.2.2　查看连接组的使用情况

1．通过 SQL Monitor Report 查看

可以执行以下 SQL 语句生成 HTML 报告，注意需要输入对应的 SQL_ID，代码如下：

```
SET TRIMSPOOL ON
SET TRIM ON
SET PAGES 0
SET LINESIZE 1000
SET LONG 1000000
SET LONGCHUNKSIZE 1000000
SPOOL /tmp/jg_report.htm
SELECT DBMS_SQL_MONITOR.REPORT_SQL_MONITOR（
       sql_id      => :b_sqlid,
       report_level => 'ALL',
       TYPE        => 'active')
FROM  DUAL;
SPOOL OFF
```

用浏览器打开/tmp/jg_report.htm 文件，如图 6-1 所示。

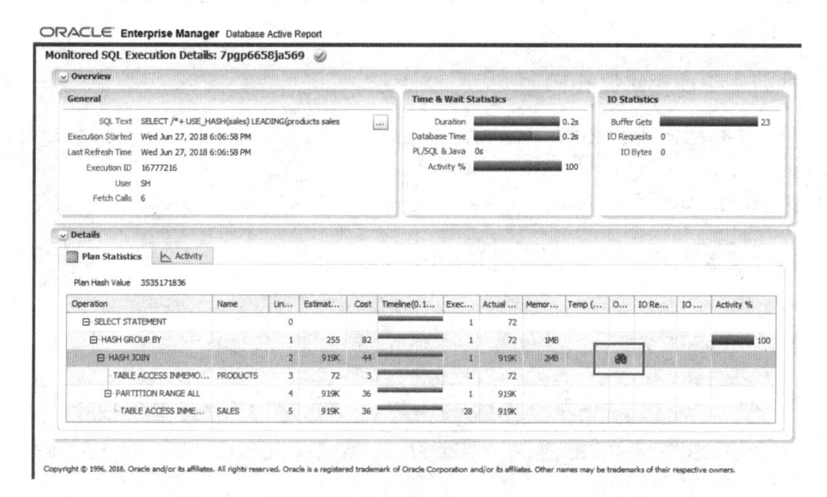

图 6-1　用浏览器打开的/tmp/jg_report.htm 文件

单击图 6-1 方框中的望远镜图标，会打开如图 6-2 所示的详细信息对话框，最后一行显示一个连接组被使用。

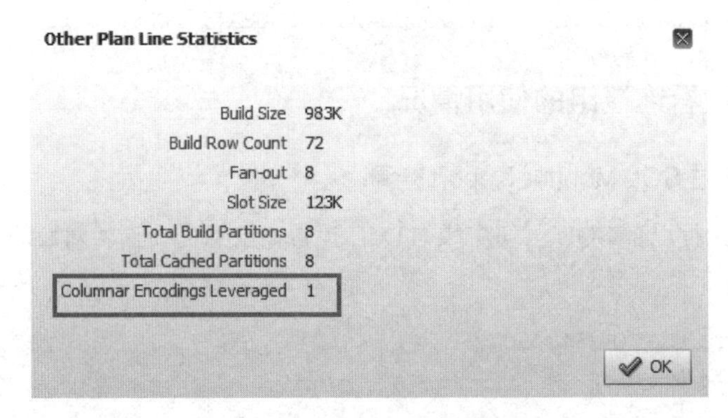

图 6-2　详细信息对话框

2. 通过命令行查看

除可以用 SQL Monitor Report 查看连接组的使用情况外，还可以直接使用命令行查看查询是否使用了连接组。

定义一个变量，用来存放 SQL_ID，代码如下：

```
SQL>VAR b_sqlid VARCHAR2 (13)
```

对表启用 INMEMORY，代码如下：

```
SQL>ALTER TABLE sales INMEMORY MEMCOMPRESS FOR QUERY;
SQL>ALTER TABLE products INMEMORY MEMCOMPRESS FOR QUERY;
```

创建一个连接组，代码如下：

```
SQL>CREATE INMEMORY JOIN GROUP cndba_class_jg (student(class_id),
class (class_id)) ;
Join group created.
```

执行全表扫描，将表存储到 IM 中，代码如下：

```
SQL>SELECT /*+ FULL (s) */ COUNT (*) FROM products s;
SQL>SELECT /*+ FULL (p) */ COUNT (*) FROM sales p;
```

将 class_id 作为连接条件执行查询操作，代码如下：

```
SQL>SELECT /*+ USE_HASH (sales) LEADING (products sales) MONITOR
*/ products.prod_id,
```

```
        products.prod_category_id, SUM (sales.amount_sold)
FROM  products, sales
WHERE  products.prod_id = sales.prod_id
GROUP BY products.prod_category_id, products.prod_id;
```

获取 SQL_ID，代码如下：

```
SQL> BEGIN
  SELECT PREV_SQL_ID
    INTO :b_sqlid
  FROM  V$SESSION
  WHERE  SID=USERENV ('SID') ;
END;
/

PL/SQL procedure successfully completed.

SQL> print b_sqlid
B_SQLID
--------------------------------------------------------------
5j2bz5m74dqwm
```

使用 DBMS_SQLTUNE.REPORT_SQL_MONITOR_XML 查看是否使用了连接组，代码如下：

```
SQL>COL row_source_id FORMAT 999
SQL>COL columnar_encoding_usage_info FORMAT A40
SQL>SELECT
  encoding_hj.rowsource_id row_source_id,
    CASE
      WHEN encoding_hj.encodings_observed IS NULL
      AND encoding_hj.encodings_leveraged IS NOT NULL
      THEN
        'join group was leveraged on ' ||
encoding_hj.encodings_leveraged || ' processes'
      ELSE
        'join group was NOT leveraged'
    END columnar_encoding_usage_info
FROM
   (SELECT DBMS_SQLTUNE.REPORT_SQL_MONITOR_XML
(session_id=>-1,sql_id=>:b_sqlid) .
          EXTRACT (q'#//operation[@name='HASH JOIN' and
@parent_id]#') xmldata
     FROM  DUAL
    ) hj_operation_data,
```

```
XMLTABLE ('/operation'
  PASSING hj_operation_data.xmldata
  COLUMNS
   "ROWSOURCE_ID"      NUMBER PATH '@id',
   "ENCODINGS_LEVERAGED" NUMBER PATH 'rwsstats/stat[@id="9"]',
   "ENCODINGS_OBSERVED"  NUMBER PATH 'rwsstats/stat[@id="10"]'
   ) encoding_hj;
```

结果如下，显示连接组已经被使用了。

```
ROW_SOURCE_ID COLUMNAR_ENCODING_USAGE_INFO
------------- ----------------------------------------
        2 join group was leveraged on 1 processes
```

6.3 优化聚合操作

从 Oracle 12.1.0.2 开始，IM 聚合支持在扫描时使用聚合查询。KEY VECTOR 和 VECTOR GROUP BY 操作使用高效数组进行连接和聚合。优化器根据成本为 GROUP BY 操作选择 VECTOR GROUP BY，但优化器不会为 GROUP BY ROLLUP、GROUPING SETS 和 CUBE 操作选择 VECTOR GROUP BY 聚合。

注意，IM 聚合也被称为 VECTOR 聚合和 VECTOR GROUP BY 聚合。IM 聚合需要启用 IM，但是 IM 聚合不要求其涉及的表也要存储到 IM 列式存储中。

IM 聚合通过预处理小表来加速在大表上的执行速度。在 Oracle 12c 之前，GROUP BY 操作只有 HASH 和 SORT。VECTOR GROUP BY 是一种额外的基于成本的转换，它将维度表和事实表之间的连接转换为过滤条件，Oracle 可以通过该过滤条件来扫描事实表，连接关键向量，类似于 Bloom filter。

IM 聚合使矢量连接和 GROUP BY 操作能够在扫描大型表时进行。因此，这些操作在扫描时聚合，不需要等待表扫描和连接操作完成。IM 聚合可以优化 CPU 使用率，尤其是 CPU 高速缓存。

IM 聚合提高了将相对较小的表和较大的事实表进行连接查询的性能，并将事实表中的数据聚合在一起。行存储的普通表和 IM 列式存储的表都可以从 IM 聚合中受益。

以下情况不适用 IM 聚合：

- 两个极大表之间的连接；
- 超过 20 亿条数据的维度表；

- 数据库没有足够的内存。

6.4 优化 IM 列式存储的重新填充

6.4.1 IM 列式存储的重新填充

自动刷新 IM 列式存储中被修改的数据操作被称为重新填充(Repopulation)。IM 列式存储会定期刷新其中被修改的对象，用户可以通过 DBMS_INMEMORY 包来控制刷新方式和频率等。

1. 行修改和事务日志

IMCU 是一种只读结构，表在执行 DML 中的操作时并不会修改 IMCU 中的数据。在事物日志中，SMU（Snapshot Metadata Unit）与每个 IMCU 修改关联。当查询访问数据并发现修改的行时，它可以从事务日志中获取相应的 RowID，然后从缓冲区高速缓存中查询修改的行。

随着修改次数的增加，SMU 的大小及必须从事务日志或数据库缓冲区高速缓存中获取的数据量也会增加。为了避免通过访问日志降低查询性能，后台进程会自动重新填充修改后的对象。

2. 自动重新填充

当 IM 列式存储中的对象执行 DML 中的操作时，数据库会自动重新填充这些对象。

自动重新填充采取以下形式。

- 基于阈值的重新填充：取决于 IMCU 的事务日记中陈旧数据量的百分比，超过阈值就会触发重新填充。
- 逐进的重新填充：通过定期刷新列式存储数据，即使尚未达到陈旧数据量的阈值的数据，也会被填充。

在自动重新填充期间，传统的访问方式仍然可以使用。可以从缓冲区缓存或磁盘读取数据。最重要的是，IM 列式存储在事务上始终与磁盘上的数据保持一致。

3. 手动执行外部表的重新填充

外部表不支持自动重新填充。IM 列式存储管理外部表与管理内部表有所不同。外部表是只读的，不会有 DML 操作，因此不依赖事务日志。正是由于这个原因，Oracle 不会自动重新填充外部表，但是可以使用 DBMS_INMEMORY.REPOPULATE 手动刷新外部表。只有当外部表完全存储到 IM 列式存储中时，才可以进行外部表的 IM 扫描。

在 DBMS_INMEMORY.REPOPULATE 中可使用参数 FORCE，它有以下两个值。

- FALSE：默认值，只会重新填充包含被修改行所在的 IMCU。
- TRUE：Oracle 会先删除 IM 中所有的外部表，然后重建。很明显，这种方式代价很大。

举一个例子，假设 IMCU 1 包含 1~10 000 行数据，IMCU 2 包含 10 001~20 000 行数据，如果一个 SQL 语句修改了第 10 005 行数据且参数 FORCE=FALSE，那么重新填充 IMCU 2 即可；如果 FORCE=TRUE，那么 IMCU 1 和 IMCU 2 都需要重新填充。

重新填充的语法如下：

```
SQL>EXEC DBMS_INMEMORY.REPOPULATE（'CNDBA', 'ADMIN_EXT_TABLES',
'',TRUE）;
```

6.4.2 重新填充 IM 列式存储的时间

数据库根据内部算法自动重新填充 IM 列式存储，也可以手动禁用重新填充。自动重新填充会一直检查陈旧的日志并使用双份缓存。重新填充有以下两种不同的触发条件。

- 基于阈值的重新填充：当事务日志中记录的更改数量达到内部陈旧记录数的阈值时，数据库将重新填充 IMCU。当 INMEMORY_MAX_POPULATE_ SERVERS 初始化参数被设置为非 0 时，会自动触发基于阈值的重新填充。
- 逐进的重新填充（Trickle repopulation）：IMCO（In-Memory Coordinator）后台进程定期检查是否存在陈旧的数据，然后将 IMCU 添加到重新填充的队列中。这种机制不管陈旧数据的比例是否达到规定的陈旧数据比例的阈值。INMEMORY_TRICKLE_REPOPULATE_ SERVERS_PERCENT 初始化参数用于限制逐进的重新填充的后台进程的数量。如果将此初始化参数设置为 0，则禁用逐进的重新填充。

查看参数设置，代码如下：

```
SQL>SHOW PARAMETER POPULATE_SERVERS
NAME                                          TYPE       VALUE
--------------------------------------        ---------- ----------
inmemory_max_populate_servers                  integer    1
inmemory_trickle_repopulate_servers_percent    integer    1
```

上面代码中的两个参数值表示只会使用 inmemory_max_populate_servers×inmemory_trickle_repopulate_servers_percent×100%的 CPU 资源来重新填充，也就是 1×1×100%=1%的 CPU 资源。

逐进的重新填充类似于 Java 垃圾收集，其运行机制如下。

（1）IMCO 开始运行。

（2）IMCO 查看是否有填充的任务需要执行，包括与 IMCU 关联的事务日志中是否存在陈旧的数据。

（3）如果 IMCO 发现有陈旧的数据，那么它会触发 Space Management Worker Process（Wnnn）进程来创建一个新的 IMCU。在 IMCU 创建期间，Oracle 会在事务日志中记录被修改行的 RowID。

（4）IMCU 会睡眠两分钟，然后继续运行，如此循环下去。

6.4.3　影响重新填充的因素

重新填充是 Oracle 数据库内部根据特定的算法来实现的，受以下几个因素的影响。

- DML 中修改数据量的多少：随着修改数据量的不断增加，IM 中陈旧数据的比例不断增大，事务日志也在不断增加，需要使用更多的缓冲区缓存来满足查询的需要。
- DML 中操作的类型：通常插入数据比删除和更新数据拥有更小的性能开销，因为通常会将数据插入到新的数据块中。
- 数据库内被修改行的位置：修改一个数据块或一个分区中的数据比修改分开存储在整个表中的数据的影响更小。
- INMEMORY 中对象的压缩级别：压缩比例越大，在解压缩时就越消耗 CPU 资源。
- 工作进程的数量：更多的进程数量会加快重新填充的速度。

第 7 章

高可用和 IM 列式存储

7.1 IM FastStart

在启用 IM 列式存储后，也可以启用 IM FastStart。IM FastStart 可以定期将 IM 中的 IMCU 数据副本保存到磁盘上，以便在实例重新启动期间更快地重新填充数据。如果数据库在关闭后又重新打开，则数据库将从 FastStart 区域读取数据，然后将其填充到 IM 列式存储中，确保所有事务的一致性。另外需要注意的是，IM FastStart 不支持 DataGuard 的备库。

7.1.1 IM FastStart 原理

FastStart 区域是 IM FastStart 存储和管理 IM 对象数据的指定表空间。FastStart 表空间由 Oracle 自动管理，无需 DBA 干预。

每个 PDB 或 Non-CDB 只允许有一个 FastStart 区域和一个指定的 FastStart 表空间。当指定了 IM FastStart 表空间后，该表空间不能被更改或删除。在 Oracle RAC 中，所有节点共享 FastStart 数据。

使用 DBMS_INMEMORY_ADMIN.FASTSTART_ENABLE 存储过程设置 FastStart 表空间，Space Management Worker Processes （Wnnn）进程会创建一个名为 SYSDBinstance_name_LOBSEG$的空 SecureFiles LOB。

Oracle 自动管理 FastStart 区域，具体如下。

- 只要发生对象的填充或重新填充，Oracle 就会将其写入 FastStart 区域。
- 如果在一个段上定义了 ADO 策略，那么 Oracle 也会根据策略中的规则管理 FastStart 区域中的段。

- 如果 IM 中的对象被禁用了 IM 存储，那么该对象也会从 FastStart 区域中删除。
- 如果 FastStart 表空间的空间不足，则 Oracle 将使用内部算法删除最久的段数据，然后继续写入 FastStart 区域。如果没有剩余空间，则 Oracle 会停止写入 FastStart 区域。

7.1.2　启用 IM FastStart

使用 DBMS_INMEMORY_ADMIN.FASTSTART_ENABLE 存储过程为 FastStart 区域指定表空间，可以设置 FastStart 区域创建的 LOB 为 LOGGING 模式。如果将 nologging 参数设置为 TRUE（默认），则数据库使用 NOLOGGING 选项创建 FastStart LOB；如果将 nologging 参数设置为 FALSE，则数据库使用 LOGGING 选项创建 FastStart LOB。

创建 FastStart 区域，必须满足以下条件。

- 分配给 FastStart 区域的表空间必须要存在。
- 分配给 FastStart 区域的表空间必须有足够的空间，Oracle 建议是 IM 空间大小（INMEMORY_SIZE 参数值）的两倍。在分配 FastStart 区域之前，表空间中不能包含其他数据。

启用 IM FastStart 的具体操作步骤如下。

（1）创建表空间，代码如下：

```
    SQL> create tablespace fs_cndba datafile
'/opt/oracle/oradata/ORCLCDB/dave/FS01.dbf' size 400m;
    Tablespace created.
```

（2）启用 IM FastStart，并将表空间分配给 FastStart 区域。对于 FastStart LOB，启用默认的 NOLIGGING 模式，代码如下：

```
    SQL>EXEC DBMS_INMEMORY_ADMIN.FASTSTART_ENABLE（'FS_CNDBA'）;
        PL/SQL procedure successfully completed.
```

（3）查看 FastStart 区域的状态和大小，代码如下：

```
    SQL>COL TABLESPACE_NAME FORMAT a15
    SQL>SELECT TABLESPACE_NAME, STATUS,（ (ALLOCATED_SIZE/1024) /
1024 ) AS ALLOC_MB,（ (USED_SIZE/1024) / 1024 ) AS USED_MB FROM
V$INMEMORY_FASTSTART_AREA;
```

```
TABLESPACE_NAME   STATUS    ALLOC_MB    USED_MB
------------- -------------- ------------- ----------------
FS_CNDBA           ENABLE     400       1.1875
```

（4）查看 FastStart LOB 的 NOLIGGING 模式，代码如下：

```
SQL>COL SEGMENT_NAME FORMAT a20
SQL>SELECT SEGMENT_NAME, LOGGING FROM  DBA_LOBS WHERE
TABLESPACE_NAME = 'FS_CNDBA';
SEGMENT_NAME     LOGGING
---------------------------- ----------------
SYSDBIMFS_LOBSEG$    NO
```

（5）对表启用 IM 并存储到 IM 列式存储中，代码如下：

```
SQL> conn cndba/cndba@cndba
Connected.
SQL> create table ustc as select * from dba_objects;
Table created.
SQL> create table cndba as select * from dba_objects;
Table created.
SQL> alter table cndba inmemory;
Table altered.
SQL> alter table ustc inmemory;
Table altered.
```

（6）进行全表扫描或手动执行存储过程，将表存储到 IM 中，代码如下：

```
SQL>SELECT /*+ FULL (s) NO_PARALLEL (s) */ COUNT (*) FROM cndba
s;
SQL>SELECT /*+ FULL (p) NO_PARALLEL (p) */ COUNT (*) FROM ustc p;
或
SQL> EXEC DBMS_INMEMORY.POPULATE ('CNDBA', 'CNDBA');
PL/SQL procedure successfully completed.
SQL> EXEC DBMS_INMEMORY.POPULATE ('CNDBA', 'USTC');
PL/SQL procedure successfully completed.
```

（7）再次查看 FastStart 区域的大小，代码如下：

```
SQL>SELECT TABLESPACE_NAME, STATUS, ( (ALLOCATED_SIZE/1024) /
1024 ) AS ALLOC_MB, ( (USED_SIZE/1024) / 1024 ) AS USED_MB FROM
V$INMEMORY_FASTSTART_AREA;
TABLESPACE_NAME STATUS              ALLOC_MB   USED_MB
--------------- -------------------- ---------- ----------
FS_CNDBAENABLE                       400    1.1875
```

（8）FastStart 区域的大小没有变化，多进行几次插入操作，代码如下：

```
SQL> insert into ustc select * from ustc;
291944 rows created.
SQL> /
583888 rows created.
SQL> /
1167776 rows created.
SQL> commit;
Commit complete.
SQL> select count (1)  from ustc;
  COUNT (1)
----------
   2335552
SQL> SELECT TABLESPACE_NAME, STATUS, ( (ALLOCATED_SIZE/1024) /
1024 ) AS ALLOC_MB, ( (USED_SIZE/1024) / 1024 ) AS USED_MB FROM
V$INMEMORY_FASTSTART_AREA;
TABLESPACE_NAME STATUS               ALLOC_MB   USED_MB
--------------- -------------------- ---------- ----------
FS_CNDBAENABLE                        400       17.1875
```

7.1.3　迁移 FastStart 区域到其他表空间

可以使用 DBMS_INMEMORY_ADMIN.FASTSTART_MIGRATE_ STORAGE 存储过程将 FastStart 区域迁移到其他表空间。注意，只能分配一个表空间给 FastStart 区域。迁移 FastStart 区域到其他表空间的具体操作步骤如下。

（1）查看当前 FastStart 的表空间，代码如下：

```
SQL>COL TABLESPACE_NAME FORMAT a15
SQL>SELECT TABLESPACE_NAME, STATUS FROM
V$INMEMORY_FASTSTART_AREA;
TABLESPACE_NAME  STATUS
--------------  -----------
FS_CNDBA         ENABLE
```

（2）创建新的表空间，代码如下：

```
SQL> create tablespace fs_ustc datafile
'/opt/oracle/oradata/ORCLCDB/dave/ustc01.dbf' size 400m;
Tablespace created.
```

（3）迁移到新的表空间，代码如下：

```
    SQL> EXEC DBMS_INMEMORY_ADMIN.FASTSTART_MIGRATE_ STORAGE
('fs_ustc') ;
    PL/SQL procedure successfully completed.
```

（4）再次查看当前 FastStart 的表空间，代码如下：

```
    SQL>COL TABLESPACE_NAME FORMAT a15
    SQL>SELECT TABLESPACE_NAME, STATUS FROM
V$INMEMORY_FASTSTART_AREA;
    TABLESPACE_NAME  STATUS
    --------------- -----------
    FS_USTC    ENABLE
```

7.1.4　禁用 FastStart

可以使用 DBMS_INMEMORY_ADMIN.FASTSTART_DISABLE 存储过程来禁用 FastStart，具体操作步骤如下。

（1）查看当前 FastStart 区域的状态，代码如下：

```
    SQL>SELECT TABLESPACE_NAME, STATUS FROM
V$INMEMORY_FASTSTART_AREA;
    TABLESPACE_NAME  STATUS
    --------------- -----------
    FS_USTC    ENABLE
```

（2）禁用 FastStart，代码如下：

```
    SQL> EXEC DBMS_INMEMORY_ADMIN.FASTSTART_DISABLE;
    PL/SQL procedure successfully completed.
```

（3）再次查看当前 FastStart 区域的状态，代码如下：

```
    SQL> SELECT TABLESPACE_NAME, STATUS FROM
V$INMEMORY_FASTSTART_AREA;
    TABLESPACE_NAME  STATUS
    ---------------- -----------
    INVALID_TABLESPACE   DISABLE
```

（4）删除表空间，代码如下：

```
    SQL> drop tablespace fs_ustc including contents and datafiles;
    Tablespace dropped.
```

7.2　在 RAC 中部署 IM 列式存储

7.2.1　RAC 中的 IM 列式存储

每个 Oracle RAC 节点都有自己的 IM 列式存储。默认情况下，对象将存储在集群中所有节点的 IM 列式存储上。Oracle 建议 RAC 每个节点的 IM 列式存储的大小相同。

可以在每个节点上存储不同的对象，或者在集群中的所有 IM 列式存储上分布存储较大的对象，但是只有在 Oracle Engineered 系统（如 Exadata）上才能够在每个节点的 IM 中存储相同的完整对象。对象在整个 IM 列式存储中的分布方式由两个参数决定：DISTRIBUTE 和 DUPLICATE。

在 Oracle RAC 环境中，启用了 INMEMORY 的对象会自动分布在集群中的 IM 列式存储上。也可以使用 DISTRIBUTE 子句来指定对象在集群中的分布方式。默认情况下，使用的分区类型（如果有）将决定对象的分布方式。如果对象没有分区，那么将通过 RowID 的范围进行分配。可以指定 DISTRIBUTE 子句来覆盖默认行为。

1．DISTRIBUTE

INMEMORY 的 DISTRIBUTE 子句控制 IM 列式存储中的表数据如何分布在 Oracle RAC 的各个节点中。

当 DISTRIBUTE 被设置为 AUTO 时，Oracle RAC 实例会自动分配数据。在填充段时，Space Management Slave Processes （Wnnn）进程会尝试在每个实例上放入等量的数据。分配方式取决于访问模式和对象大小。也可以手动指定数据库如何在实例间分配分区、子分区或 RowID 的范围。

数据的平均分布对于性能非常重要。在相同大小的数据上执行并行查询，可以在最短的时间内完成。如果数据分布过于倾斜，则查询时间取决于最长查询的那个节点的时间。

如果 Oracle RAC 中的一个节点挂掉，那么 IMCU 将不可用。因此，需要存储在不可访问的 IMCU 中的数据的查询必须从其他位置读取数据，如数据库缓冲区高速缓存、闪存、磁盘或其他 IM 列式存储中的镜像 IMCU。DBA_TABLES 视图中的 INMEMORY_DISTRIBUTE 列显示了 IMCU 是如何分布的。

（1）按（子）分区分布

可以使用 DISTRIBUTE BY PARTITION|SUBPARTITION 将分区中的数据分布存储到 RAC 的不同节点上，这个数据分布方式对于 HASH（散列）分区特别适合。

如图 7-1 所示显示了表 CNDBAS 根据 USER_ID 进行散列分区，然后 Oracle 将不同的分区分别存储到不同的节点上。

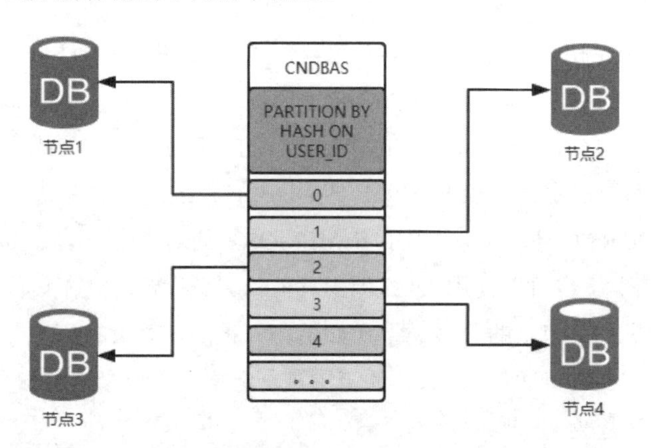

图 7-1　按分区分布

按分区分布的代码如下：

```
CREATE TABLE cndba.ustc (
PK_ID NUMBER (30) NOT NULL,
ADD_DATE_TIME       DATE,
CONSTRAINT PK_T_TEST PRIMARY KEY (PK_ID)
)
PARTITION BY RANGE (add_date_time)
(PARTITION cndba_test _2016_less VALUES LESS THAN (TO_DATE
('2016-01-01 00:00:00','yyyy-mm-ddhh24:mi:ss')) ,
  PARTITION cndba_test _2017 VALUES LESS THAN (TO_DATE('2017-01-01
00:00:00','yyyy-mm-ddhh24:mi:ss')) ,
  PARTITION cndba_test _2018 VALUES LESS THAN (TO_DATE('2018-01-01
00:00:00','yyyy-mm-dd hh24:mi:ss')))
INMEMORY DISTRIBUTE BY PARTITION DUPLICATE ALL;
```

（2）按 RowID 的范围分布

可以使用 DISTRIBUTE BY ROWID RANGE 将指定范围内的 RowID 存储到不同的节点上，如图 7-2 所示，将 RowID 1~100 存储到节点 1，将 RowID

101~200 存储到节点 2，以此类推。

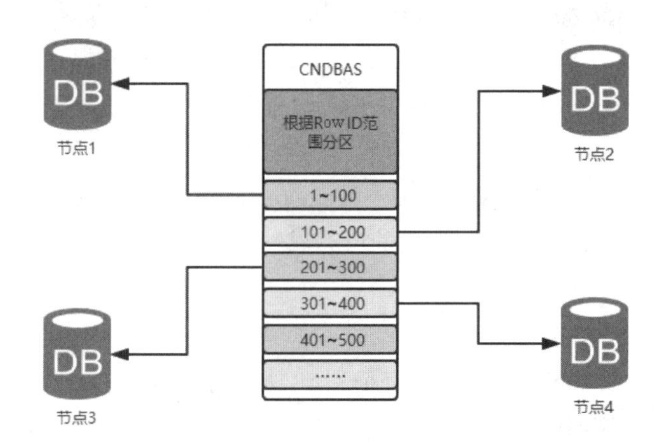

图 7-2　按 RowID 的范围分布

RowID 分布技术非常适合非分区表。但是，如果分区策略导致数据严重倾斜，例如一个分区的大小比其他分区的大小大得多，那么 Oracle 建议通过手动指定 DISTRIBUTE BY ROWID RANGE 的方式来覆盖默认分配方式（BY PARTITION）。

按 RowID 的范围分布的语法如下：

CREATE TABLE ustc（id number） INMEMORY DISTRIBUTE BY ROWID RANGE NO DUPLICATE;

2. Oracle RAC 中列式存储数据的复制

可以通过 DUPLICATE 和 NO DUPLICATE 来指定数据在 RAC 不同节点之间的分布方式。目前只有 Exadata 上的 RAC 才能使用 DUPLICATE。只要不是 Oracle Engineered System 环境，即便指定了 DUPLICATE，Oracle 也会当作 NO DUPLICATE 来处理。

（1）DUPLICATE

DUPLICATE 控制着 Oracle 在实例之间复制数据的方式，也就是将 RAC 中的一个对象在 RAC 的第二个实例的 IM 中保存一份完整的副本。因此，同一个数据对象会在两个 Oracle RAC 实例中进行填充，而且仅有两个实例（一主一备）。如果主数据的实例出现故障，那么查询可以通过数据副本获取数据。

（2）DUPLICATE ALL

DUPLICATE ALL 和 DUPLICATE 唯一的不同是，DUPLICATE ALL 支持更多的副本，相当于 ASM 磁盘组有更高的冗余级别。DUPLICATE ALL 支持在 RAC 的所有节点上都保存副本，所以只要有一个节点正常工作，就不会影响查询。

如果指定了 DUPLICATE ALL，那么 Oracle 默认使用 DISTRIBUTE AUTO 而忽略 inmemory_distribute 指定的所有属性，如:DISTRIBUTE BY ROWID RANGE。

（3）NO DUPLICATE

NO DUPLICATE 就是不保存任何副本，相当于磁盘组的外部冗余。如果存储数据的节点不可用，那么查询就无法从 IM 中检索数据，只能从磁盘或高速缓存区获取数据，从而影响查询的性能。如果存储数据的节点在一段时间内仍无法使用，并且 IM 中还有空闲空间，那么 Oracle 会在集群中的其他节点上填充缺少的数据（故障节点上的数据）。

（4）具体示例

创建测试表，在任意一个节点上创建一个表，可以指定压缩方式，否则使用默认的属性，代码如下：

```
SQL> show pdbs
   CON_ID CON_NAME              OPEN MODE  RESTRICTED
---------- ------------------------------ ---------- ----------
     2 PDB$SEED             READ ONLY  NO
     3 DAVE             READ WRITE NO
SQL> alter session set container=dave;
Session altered.
SQL> create user cndba identified by cndba;
User created.
SQL> grant connect,resource,dba to cndba;
Grant succeeded.

SQL>create table cndba.ustc (
pk_id number(30) not null,
add_date_time      DATE,
constraint PK_CNDBA_TEST primary key (pk_id)
)
PARTITION BY RANGE (add_date_time)
(PARTITION CNDBA_TEST_2016_less VALUES LESS THAN
```

```
    (TO_DATE ('2016-01-01 00:00:00','yyyy-mm-ddhh24:mi:ss')) ,
    PARTITION CNDBA_TEST_2017 VALUES LESS THAN (TO_DATE('2017-01-01
00:00:00','yyyy-mm-ddhh24:mi:ss')) ,
    PARTITION CNDBA_TEST_2018 VALUES LESS THAN (TO_DATE('2018-01-01
00:00:00','yyyy-mm-dd hh24:mi:ss')))
  INMEMORY DISTRIBUTE BY PARTITION DUPLICATE ALL;
```

插入数据，代码如下：

```
SQL>declare
  i   int := 1;
  year VARCHAR2 (20) ;
begin
  loop
    year := CASE mod (i, 3)
           WHEN 0 THEN
            '2015-01-14 12:00:00'
           WHEN 1 THEN
            '2016-01-14 12:00:00'
           ELSE
            '2017-01-14 12:00:00'
          END;
         insert into cndba.ustc values (i, to_date (year,
'yyyy-mm-dd hh24:mi:ss')) ;
    exit when i= 2000;
    i := i + 1;
  end loop;
end;
/
SQL> commit;
Commit complete.
```

将表填充到 IM 中，手动执行或进行一次全表扫描查询都可以将表存储到 IM 中，代码如下：

```
SQL> EXEC DBMS_INMEMORY.POPULATE ('CNDBA', 'USTC') ;
```

查看 IM 中的信息，可以看到分区表 USTC 中的不同分区被存储到不同节点的 IM 中，代码如下：

```
SQL>COL SEGMENT_NAME FOR A20
SQL>COL PARTITION_NAME FOR A30
SQL>SET LINE 200
SQL>SELECT INST_ID,SEGMENT_NAME,PARTITION_NAME,
```

```
POPULATE_STATUS,INMEMORY_SIZE,INMEMORY_DISTRIBUTE,
    INMEMORY_DUPLICATE FROM GV$IM_SEGMENTS WHERE SEGMENT_NAME =
'USTC';
    INST_ID SEGMENT_NAME       PARTITION_NAME
POPULATE_STATUS          INMEMORY_SIZE INMEMORY_DISTRIBUTE
INMEMORY_DUPLICATE
    ------- ------------------------------ -------------
---------------------- --------------------------
     1 USTC      CNDBA_TEST_2018       COMPLETED          1310720
BY PARTITION          DUPLICATE ALL
     1 USTC      CNDBA_TEST_2016_LESS  COMPLETED          1310720
BY PARTITION          DUPLICATE ALL
     2 USTC      CNDBA_TEST_2017       COMPLETED          1310720
BY PARTITION          DUPLICATE ALL
```

关闭节点 2 的实例，因为分区 CNDBA_TEST_2017 是存储在节点 2 的 IM 中的，所以关闭节点 2，然后查询该分区中的数据，代码如下：

```
SQL> select instance_number from v$instance;

INSTANCE_NUMBER
---------------
              2

SQL> shutdown abort
ORACLE instance shut down.
```

查看 IM 中的信息，可以看到分区 CNDBA_TEST_2017 已经不在 IM 中了，代码如下。虽然表是通过 DUPLICATE 方式将数据分开存储到集群中的，但是由于不是 Exadata 环境，所以 Oracle 采用 NO DUPLICATE 方式存储表数据，也就是没有副本。如果存储该数据的节点出现故障，那么相应的数据也就无法从 IM 中查询，只能从内存或硬盘中读取。

```
    SQL>SELECT INST_ID,SEGMENT_NAME,PARTITION_NAME,
POPULATE_STATUS,INMEMORY_SIZE,INMEMORY_DISTRIBUTE,
    INMEMORY_DUPLICATE FROM GV$IM_SEGMENTS WHERE SEGMENT_NAME =
'USTC';
       INST_ID SEGMENT_NAME   PARTITION_NAME
POPULATE_STATUS         INMEMORY_SIZE INMEMORY_DISTRIBUTE
INMEMORY_DUPLICATE
    ---------- -------------------- ------------------------------
-------------------------- -------------
```

```
------------------------------ --------------------------
    1 USTC      CNDBA_TEST_2018       COMPLETED          1310720
BY PARTITION          DUPLICATE ALL
    1 USTC      CNDBA_TEST_2016_LESS      COMPLETED       1310720
BY PARTITION          DUPLICATE ALL
```

如果实例 2 关闭的时间过长，那么 Oracle 会自动将存储在实例 2 上 IM 中的数据重新填充到其他节点的 IM 中（本环境只有两个节点，会填充到节点 1 的 IM 中）。

手动将表重新填充到 IM 中，在节点 1 上手动执行填充表操作，Oracle 就会将表存储到节点 1 的 IM 中，代码如下：

```
SQL> EXEC DBMS_INMEMORY.POPULATE('CNDBA', 'USTC');
PL/SQL procedure successfully completed.
```

查看 IM 中的信息，此时可以看到分区表 USTC 的所有分区都存储在节点 1 的 IM 中，代码如下：

```
    SQL>SELECT INST_ID,SEGMENT_NAME,PARTITION_NAME,
POPULATE_STATUS,INMEMORY_SIZE,INMEMORY_DISTRIBUTE,
    INMEMORY_DUPLICATE FROM GV$IM_SEGMENTS WHERE SEGMENT_NAME =
'USTC';
      INST_ID SEGMENT_NAME   PARTITION_NAME
POPULATE_STATUS        INMEMORY_SIZE INMEMORY_DISTRIBUTE
INMEMORY_DUPLICATE
    ---------- -------------------- ------------------------------
------------------------------ --------------------------
     1 USTC      CNDBA_TEST_2018       COMPLETED          1310720
BY PARTITION          DUPLICATE ALL
     1 USTC      CNDBA_TEST_2017       COMPLETED          1310720
BY PARTITION          DUPLICATE ALL
     1 USTC      CNDBA_TEST_2016_LESS  COMPLETED          1310720
BY PARTITION          DUPLICATE ALL
```

可以看到，对于启用了 DUPLICATE ALL 的表，理论上 Oracle 会在所有节点的 IM 中保存该表的数据，但是仅对 Exadata 环境有效，其他环境 Oracle 都将其当作 NO DUPLICATE 来处理，也就是不存储数据副本。

7.2.2 RAC 中的 FastStart

RAC 中的 FastStart 的操作和普通单实例中的 FastStart 的配置、迁移等相关操作没有任何区别。需要注意的是，不管对象在 IM 中保存了几份副本，FastStart 只会存储其中的一份副本。例如，在一个有 3 个节点的 RAC 集群中，一个表启用了 DUPLICATE ALL，也就意味着该表在 IM 中有 3 份副本，但是 FastStart 只会保存其中的一份副本。

7.3 在 ADG 中部署 IM 列式存储

Oracle 从 12.2.0.1 版本开始支持 ADG 中的 IM，但是只有 Oracle Engineered Systems（Exadata）和 Oracle Cloud Platform 才支持 ADG 中的 IM。

要配置 ADG 中的 IM，必须满足以下条件。

- 备库必须运行在 Oracle Engineered Systems 或 Oracle Cloud Platform 上。
- COMPATIBLE 参数必须设置为 12.2.0 及以上。
- 为了在主、备库上填充不同的对象，需要配置合适的服务。

第 8 章

Oracle ASM 概述

Oracle 10g 中引入了 ASM。ASM 是一个数据文件的卷管理器和文件系统，支持单实例和 RAC，用于替代传统的卷管理器、文件系统、裸设备。Oracle ASM 使用磁盘组来存储数据文件，并且存储在磁盘组中的数据是均匀分布的，可以消除热点。Oracle 官方也宣称对磁盘组的 I/O 性能进行了特殊的优化，可以与裸设备的性能相媲美。

Oracle ASM 涉及的知识点有 ASM 实例、磁盘组、故障组、磁盘、AU（Allocation Units）和 ASM 文件等。在学习 ASM 的特性时，需要重点掌握磁盘组的各种操作，并理解这些操作的含义。

8.1 ASM 实例

ASM 实例在本质上和数据库实例是一样的，它们是相互独立运行的，都有自己的后台进程和内存区域。因为 ASM 只管理很少的任务，所以只需要很小的内存，对服务器的性能影响也极小。

从下面的查询结果中可以看到，ASM 实例和 DB 实例都有自己对应的后台进程。

```
[grid@rac2 ~]$ ps -ef|grep smon
root      2941     1  1 Sep09 ?        00:15:10
/u01/app/18.3.0/grid/bin/osysmond.bin
grid      4277  4154  0 08:55 pts/0    00:00:00 grep --color=auto
smon
grid      5057     1  0 Sep09 ?        00:00:01 asm_smon_+ASM2
grid      5908     1  0 Sep09 ?        00:00:02 mdb_smon_-MGMTDB
oracle    6203     1  0 Sep09 ?        00:00:02 ora_smon_cndba2
[grid@rac2 ~]$ ps -ef|grep pmon
```

```
    grid     4390  4154  0 08:55 pts/0    00:00:00 grep --color=auto
pmon
    grid     5010     1  0 Sep09 ?       00:00:03 asm_pmon_+ASM2
    grid     5636     1  0 Sep09 ?       00:00:02 apx_pmon_+APX2
    grid     5854     1  0 Sep09 ?       00:00:03 mdb_pmon_-MGMTDB
    oracle    6129     1  0 Sep09 ?       00:00:03 ora_pmon_cndba2
    [grid@rac2 ~]$
```

ASM 实例和数据库实例分开管理，在 RAC 环境中，ASM 实例所对应的实例名是+ASM1、+ASM2，其实例只能启动到 STARTED 状态，这是一种介于 STARTUP NOMOUNT 和 STARTUP MOUNT 之间的状态，代码如下。ASM 实例主要是对磁盘组进行管理，并为数据库实例提供数据访问的服务。

```
    [grid@rac1 trace]$ sqlplus / as sysdba
    SQL*Plus: Release 18.0.0.0.0 - Production on Mon Sep 10 08:57:26
2018
    Version 18.3.0.0.0

    Copyright (c) 1982, 2018, Oracle.  All rights reserved.

    Connected to:
    Oracle Database 18c Enterprise Edition Release 18.0.0.0.0 -
Production
    Version 18.3.0.0.0

    SQL> select instance_name,status from v$instance;

    INSTANCE_NAME    STATUS
    ---------------- ------------
    +ASM1            STARTED
```

在安装 GRID 时会自动安装 ASM 实例，RAC 集群的每个节点上只能有一个 ASM 实例。如果同一个节点上有多个数据库实例，那么多个数据库实例将共享一个 ASM 实例。如果该 ASM 实例挂掉了，那么该节点上的所有数据库实例也将无法使用。注意这里 Flex ASM 除外，Oracle 从 12c 开始提供了 Flex Cluster 和 Flex ASM 功能，并且在 Oracle 18c 中默认启用该功能。在使用 Flex ASM 的情况下，如果一个节点上的 ASM 实例挂掉了，那么该节点上的数据库实例会重新连接到其他节点的 ASM 实例上继续运行。这个过程对用户透明，用户感觉不到。

8.2　ASM 磁盘组

ASM 磁盘组由磁盘组成，是 Oracle ASM 中的基本管理对象。每个磁盘组都包含管理磁盘组中的空间所需的元数据，元数据存储在磁盘组中。Oracle ASM 元数据包括以下内容：

- 磁盘与磁盘组的对应关系；
- 磁盘组中可用的空间大小；
- 磁盘组中的文件名；
- 磁盘组数据文件扩展（Extent）的位置；
- 修改元数据产生的 Redo Log；
- Oracle ADVM 卷信息。

数据库文件存储在 ASM 磁盘组中，一个 Oracle ASM 文件只能存储在一个磁盘组中。

如下代码显示了 ASM 环境中有 3 个磁盘组。

```
[grid@rac1 ~]$ asmcmd lsdg
State    Type    Rebal  Sector  Logical_Sector  Block      AU
Total_MB  Free_MB  Req_mir_free_MB  Usable_file_MB  Offline_disks
Voting_files  Name
MOUNTED  NORMAL  N       512           512   4096  4194304
61440    43440            0         21720           0          N
DATA/
MOUNTED  EXTERN  N       512           512   4096  4194304
40960    35692            0         35692           0          N
MGMT/
MOUNTED  NORMAL  N       512           512   4096  4194304
6144     5228          2048          1590           0          Y
OCR/
[grid@rac1 ~]$
```

8.3　镜像和故障组

镜像通过将磁盘上的数据复制成多个故障组来保护数据的完整性，这样不会因为一个磁盘的故障而导致数据的丢失。镜像数是由磁盘组的冗余类型决定的。在 Oracle 18c 中，磁盘组有 5 种冗余类型，如表 8-1 所示。

表 8-1 磁盘组的 5 种冗余类型

磁盘组冗余类型	支持的镜像级别	默认镜像级别
External	没有镜像	没有镜像
Normal	两份镜像、三份镜像、没有镜像	两份镜像
High	三份镜像	三份镜像
Flex	两份镜像、三份镜像、没有镜像	两份镜像（新建的磁盘组）
Extent	两份镜像、三份镜像、没有镜像	两份镜像

根据磁盘组的镜像级别可以知道：

- External 冗余类型的磁盘组，任何磁盘损坏都会造成数据丢失；
- Normal 冗余类型的磁盘组，允许一个故障组或其中的磁盘损坏且不丢失数据；
- High 冗余类型的磁盘组，允许两个故障组或其中的磁盘损坏且不丢失数据；
- Flex 冗余类型的磁盘组，最多允许两个故障组或其中的磁盘损坏且不丢失数据；
- Extent 冗余类型的磁盘组，最多允许两个故障组或其中的磁盘损坏且不丢失数据。

Extent 磁盘组具备 Flex 磁盘组的所有特性，从某种角度看，它们是一种磁盘组，所以在 Oracle 的 GUI 界面上只显示了 4 种冗余类型，如图 8-1 所示。

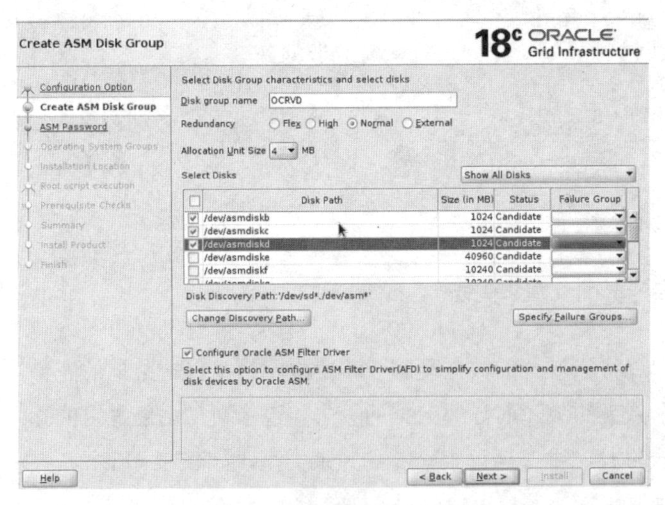

图 8-1 Oracle 的 GUI 界面

磁盘组的冗余类型可以通过 SQL 语句查询，也可以通过 asmcmd 命令查询，在 8.2 节的磁盘组查询结果中也可以看到磁盘组的冗余类型。以下 SQL 语句查询的结果包含磁盘的 MOUNT 状态、磁盘组的冗余类型和故障组等信息。

```
SQL> SELECT SUBSTR (dg.name, 1, 16) AS diskgroup,
      SUBSTR (d.name, 1, 16) AS asmdisk,
      d.mount_status,
      d.state,
      SUBSTR (d.failgroup, 1, 16) AS failgroup
  FROM V$ASM_DISKGROUP dg, V$ASM_DISK d
 WHERE dg.group_number = d.group_number;

DISKGROUP     ASMDISK  MOUNT_S STATE    FAILGROUP
---------------- ------------ ------- -------- ----------------
DATA       DATA_0000    CACHED  NORMAL   FG1
DATA       DATA_0001    CACHED  NORMAL   FG2
MGMT       MGMT1        CACHED  NORMAL   MGMT1
OCR        OCR1         CACHED  NORMAL   OCR1
OCR        OCR2         CACHED  NORMAL   OCR2
OCR        OCR3         CACHED  NORMAL   OCR3

6 rows selected.
```

当 Oracle ASM 为镜像文件分配区（Extent）时，ASM 会分配主副本和镜像副本，并且镜像副本与主副本被存储在不同的故障组的磁盘上，即使同一个故障组中所有磁盘同时损坏，也不会造成数据丢失。

在创建磁盘组时，可以为每个磁盘指定名称，如果不指定，Oracle 会根据磁盘组的名称自动命名磁盘的名称，也可以自定义故障组名称。在 Oracle 18c 之前无法修改磁盘组的冗余类型，从 Oracle 18c 开始，支持从 Normal 或 High 冗余类型修改到 Flex 冗余类型。在创建磁盘组时，如果没有指定故障组，那么 Oracle 会默认将磁盘存放到自己的故障组中。Normal 冗余类型最少需要两个故障组，High 冗余类型最少需要三个故障组，External 冗余类型不需要故障组。

可以使用如下 SQL 语句查看相关的信息，这里 FG1 和 FG2 是在创建磁盘组时指定的，OCR1、OCR2 和 OCR3 是默认的。

```
SQL> col path for a30
SQL> select
group_number,disk_number,total_mb,name,failgroup,path from
v$asm_disk order by 1;
```

```
GROUP_NUMBER DISK_NUMBER   TOTAL_MB NAME     FAILGROUP   PATH
------------ ------------------------- ---------------------
       0       0    0                             /dev/asmdiskd
       0       1    0                             /dev/asmdiskb
       0       2    0                             /dev/asmdiskh
       0       3    0                             /dev/asmdiske
       0       4    0                             /dev/asmdiskc
       1       0     30720 DATA_0000      FG1     /dev/asmdiskf
       1       1     30720 DATA_0001      FG2     /dev/asmdiskg
       2       0     40960 MGMT1          MGMT1    AFD:MGMT1
       3       0      2048 OCR1           OCR1     AFD:OCR1
       3       1      2048 OCR2           OCR2     AFD:OCR2
       3       2      2048 OCR3           OCR3     AFD:OCR3
11 rows selected.
```

8.4 AU 和 ASM 文件

AU（Allocation Units）是磁盘组的基本分配单位。扩展区（Extent）由一个或多个 AU 组成。Oracle ASM 文件由一个或多个扩展区组成。在创建磁盘组时，可以使用 AU_SIZE 参数来设置磁盘组的 AU 大小，可以是 1 MB 、2 MB 、4 MB 、8 MB 、16 MB 、32 MB 或 64 MB。较大的 AU 可以为大型的使用顺序读取的数据仓库应用程序提供更好的性能。

ASM 文件就是存储在磁盘组中的文件。每个 ASM 文件只属于一个磁盘组，数据实际上存储在磁盘组的磁盘文件中。在磁盘组中可以存储不同的文件类型，Oracle 18c 目前支持以下类型的文件：

- 控制文件；
- 数据文件、临时文件、数据文件副本；
- SPFILE 文件；
- 联机日志文件、归档日志文件、闪回日志文件；
- RMAN 备份文件；
- 恢复配置信息文件；
- Trace 信息文件；
- 数据泵转储文件。

Oracle ASM 自动生成 Oracle ASM 文件名，Oracle ASM 文件名以加号（+）开头，后面跟磁盘组名称。也可以为 Oracle ASM 文件指定别名，并为别名创建分层目录结构。

在 Oracle 11g 中，Oracle 建议 AU_SIZE 为 4MB，但实际默认大小还是 1MB，

到了 Oracle 18c，AU_SIZE 默认就是 4MB 了，代码如下：

```
    SQL> select name, allocation_unit_size/1024/1024||'M' as AU_SIZE
from v$asm_diskgroup;

    NAME                 AU_SIZE
    -------------- --------------------
    DATA                 4M
    MGMT                 4M
    OCR                  4M
```

8.5 ASM 扩展区

每个扩展区位于单独的磁盘上。扩展区由一个或多个 AU 组成。为了适应文件大小，Oracle ASM 会根据策略自动使用不同大小的扩展。

使用可变大小的扩展可以支持更大的 Oracle ASM 数据文件，从而减少数据库 SGA 的大小，并提高文件创建和操作的性能。初始扩展大小等于磁盘组 AU 的大小，根据策略会自动增加 4 倍或 16 倍的 AU 大小。

对于 AU 大小小于 4MB 的磁盘组，可变扩展大小规则如下：

- 前 20000（0~19999）个扩展，每个扩展大小等于 AU；
- 后面的 20000（20000~39999）个扩展，每个扩展大小等于 AU×4；
- 再往后面的所有扩展(40000 及后面的)，每个扩展大小等于 AU×16。

磁盘组、磁盘、文件、拓展区、AU 的关系是一对多的，它们之间的逻辑关系如图 10-2 所示。

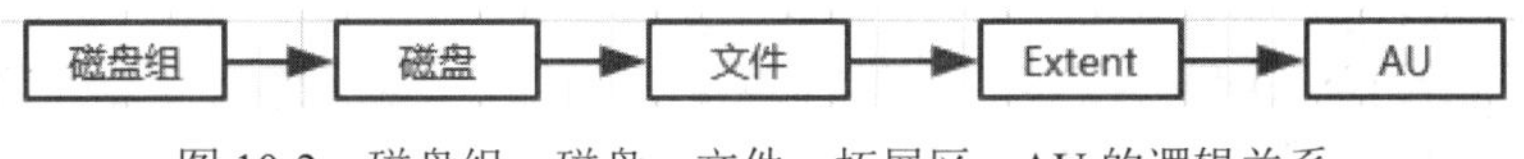

图 10-2 磁盘组、磁盘、文件、拓展区、AU 的逻辑关系

8.6 ASM 条带化

Oracle ASM 条带化有两个主要作用：平衡磁盘组中磁盘的负载和减少 I/O 延迟。

粗粒度条带化为磁盘组提供负载均衡，而细粒度条带化通过分开存储来减少 I/O 延迟。

为了存储条带化数据，Oracle ASM 将文件分成条带，并在磁盘组中的所有磁盘上均匀分布数据。细粒度条带的大小始终等于 128 KB，可以为低 I/O 操作减少 I/O 延迟。粗粒度条带的大小始终等于 AU 的大小（不是扩展区的大小）。

- 细粒度条带化：文件以 128 KB 大小的块（标记为 A、B……X）为单位进行条带化存储，从磁盘 1 中的第一个扩展区开始，接着是磁盘 2 中的第一个扩展区，然后循环遍历所有磁盘，直到整个文件被存储到磁盘中。细粒度条带化存储示意图如图 8-3 所示，条带化块先存满每个磁盘的第一个扩展区，然后填充每个磁盘的第二个扩展区，以此类推，直到整个文件被存储到磁盘中。

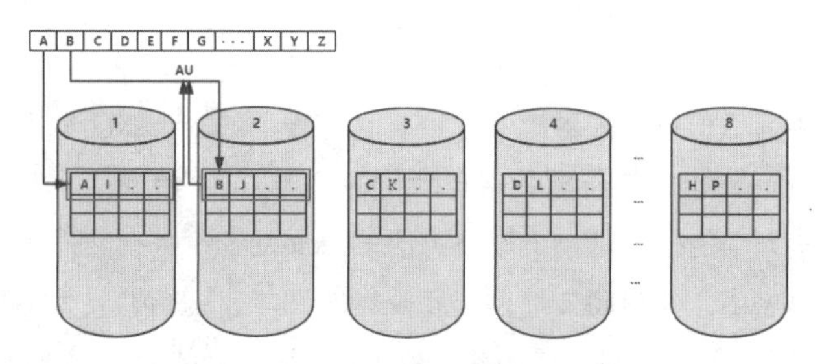

图 8-3 细粒度条带化存储示意图

- 粗粒度条带化：文件以 4 MB 大小的块（标记为 A、B……X）为单位进行条带化存储，从磁盘 1 中的第一个扩展区开始，接着是磁盘 2 中的第一个扩展区，然后循环遍历所有磁盘，直到整个文件被存储到磁盘中。对于 AU 等于扩展区大小（4MB）的前 20 000 个扩展，条带等于扩展大小和 AU 大小。对于可变的扩展，扩展由多个 AU 组成，文件条带位于扩展的 AU 中。在条带化存储到下一个扩展之前，条带块被存放在所有磁盘的第一个扩展区的 AU 中。粗粒度条带化存储示意图如图 8-4 所示。

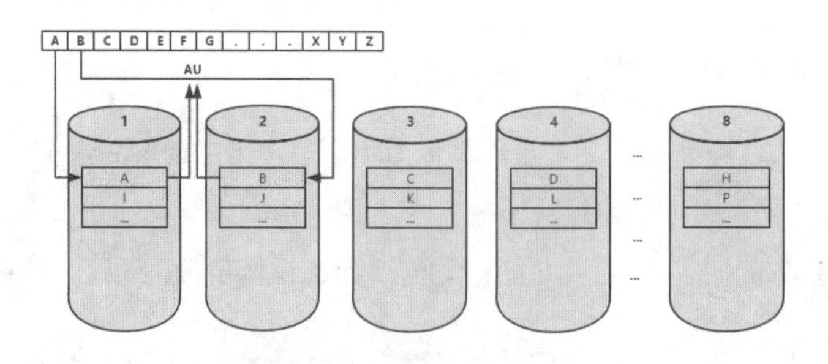

图 8-4 粗粒度条带化存储示意图

第9章

Oracle ASM 实例和磁盘组

9.1 ASM 实例管理

ASM 实例需要管理的内容不多，这里主要讲解 ASM 实例的参数。ASM 的初始化参数文件的作用和管理方法与数据库实例的初始化参数文件的作用和管理方法基本相同。

可以使用如下命令查询 ASM 参数文件的位置：

```
SQL> show parameter spfile
NAME          TYPE     VALUE
-------- ---------------------------------------------------
spfile            string
+OCR/rac/ASMPARAMETERFILE/registry.253.985999167
```

9.1.1 参数文件的维护

虽然很少对 ASM 实例的参数文件进行操作，但是 ASMCMD 工具还是支持对参数文件进行备份、复制、移动等操作。可以通过 CREATE SPFILE 命令创建新的 SPFILE 文件，也可以通过操作系统级别的命令来复制、移动 SPFILE 文件。

查看当前 SPFILE 文件所在的目录，代码如下：

```
[grid@rac1 ~]$ asmcmd spget
+OCR/rac/ASMPARAMETERFILE/registry.253.985999167
```

创建本地的 PFILE 文件，并查看参数文件的内容，代码如下：

```
SQL> create pfile from spfile;
File created.
```

```
    OCR/rac/ASMPARAMETERFILE/registry.253.985999167
    [grid@rac1 ~]$ ls
    Desktop Documents Downloads Music Pictures Public Templates
Videos
    [grid@rac1 ~]$ cd $ORACLE_HOME/dbs
    [grid@rac1 dbs]$ ls
    ab_+ASM1.dat  hc_+APX1.dat  hc_+ASM1.dat  hc_-MGMTDB.dat
init+ASM1.ora  init.ora  lk_MGMTDB
    [grid@rac1 dbs]$ cat init+ASM1.ora
    +ASM1.__oracle_base='/u01/app/grid'#ORACLE_BASE set from in
memory value
    +ASM2.__oracle_base='/u01/app/grid'#ORACLE_BASE set from in
memory value
    +ASM2._asm_max_connected_clients=6
    *.asm_diskgroups='MGMT','DATA'#Manual Mount
    *.asm_diskstring='/dev/asm*','AFD:*'
    *.asm_power_limit=1
    *.large_pool_size=12M
    *.remote_login_passwordfile='EXCLUSIVE'
    [grid@rac1 dbs]$
```

可以看到，ASM 实例的参数与 DB 实例的参数相比要少很多，这里显示的只有 5 个参数。

复制 SPFILE 文件到新的目录，代码如下：

```
    ASMCMD> spget
    +OCR/rac/ASMPARAMETERFILE/registry.253.985999167
    ASMCMD> spcopy +OCR/rac/ASMPARAMETERFILE/registry.253.985999167
+DATA
    ERROR: Unable to form destination file name.
    ERROR: Source file name has file number and incarnation in it.
Provide destination file name.
    ASMCMD-8303: invalid SPFILE 'registry.253.985999167'
    ASMCMD> spcopy +OCR/rac/ASMPARAMETERFILE/registry.253.985999167
+DATA/cndba.ora
    ASMCMD>
```

注意，第一次使用 spcopy 命令进行复制的时候会报错，提示文件无效，这是因为 ASM 中的文件使用 OMF 进行管理，文件是自动生成的，在复制时没有指定文件别名，所以复制失败。

当指定文件别名之后，后台会生成一个 OMF 的文件指向这个别名，下面

验证一下：

```
ASMCMD> ls
DATA/
MGMT/
OCR/
ASMCMD> cd data
ASMCMD> ls -l
Type             Redund  Striped  Time             Sys  Name
                                                   N    CNDBA/
ASMPARAMETERFILE  MIRROR  COARSE   SEP 10 11:00:00  N    cndba.ora
=> +DATA/rac/ASMPARAMETERFILE/REGISTRY.253.986470965
                                                   Y    rac/
```

此时再执行 spget 命令，就可以看到当前的参数文件还是原来的参数文件，代码如下：

```
ASMCMD> spget
+OCR/rac/ASMPARAMETERFILE/registry.253.985999167
ASMCMD>
```

可以使用 spcopy 命令复制参数文件，在复制的同时也可以指定-u 参数，让 ASM 在复制后直接使用新的参数文件。当然，也可以直接使用 spset 命令设置新的参数文件。

```
ASMCMD>  spcopy
+OCR/rac/ASMPARAMETERFILE/registry.253.985999167 +DATA/dave.ora
ORA-15056: additional error message
ORA-06512: at line 7
ORA-17502: ksfdcre:4 Failed to create file +DATA/dave.ora
ORA-15268: internal Oracle file +DATA.253.1 already exists.
ORA-06512: at "SYS.X$DBMS_DISKGROUP", line 635
ORA-06512: at line 3 (DBD ERROR: OCIStmtExecute)
```

这里提示出错，需要把之前的参数文件删除，再进行创建，代码如下：

```
ASMCMD> ls
CNDBA/
cndba.ora
rac/
ASMCMD> rm -rf cndba.ora
ASMCMD> spcopy -u
+OCR/rac/ASMPARAMETERFILE/registry.253.985999167 +DATA/dave.ora
ASMCMD> ls -l
```

```
Type            Redund  Striped  Time            Sys  Name
                                             N    CNDBA/
ASMPARAMETERFILE  MIRROR  COARSE   SEP 10 11:00:00  N    dave.ora =>
+DATA/rac/ASMPARAMETERFILE/REGISTRY.253.986471351
                                             Y    rac/
ASMCMD>
ASMCMD> spget
+DATA/dave.ora
```

此时 ASM 的参数文件就被修改成指定的了。下面再使用 spset 命令修改回默认的参数文件，代码如下：

```
ASMCMD> spset +OCR/rac/ASMPARAMETERFILE/registry.253.985999167
ASMCMD> spget
+OCR/rac/ASMPARAMETERFILE/registry.253.985999167
ASMCMD>
```

9.1.2　常用的 ASM 参数

在 9.1.1 小节对 ASM 参数文件进行操作时，共有 5 个参数，下面讲解其中的 3 个主要参数。

- ASM_DISKGROUPS：该参数是指定 ASM 实例启动过程中挂载的磁盘组。例如，ASM_DISKGROUPS=DATA,FRA，在启动时就会自动挂载这两个磁盘组。该参数的默认值是 NULL，支持动态修改。Oracle 12c 之前的版本 ASM 最多支持 63 个磁盘组，从 Oracle 12c 开始最多支持 511 个磁盘组。如果执行 ALTER DISKGROUP…ALL MOUNT/DISMOUNT 命令，则不受该参数值的影响。
- ASM_POWER_LIMIT：该参数表示磁盘组中磁盘再平衡（Rebalance）数据的速度，值越大，再平衡的速度越快，但是会导致更高的 I/O 开销。该参数的取值范围是 0~1024，默认值为 1。如果值为零，则表示禁用再平衡。在 Oracle 11.2.0.2 之前的版本中，可设置的最大值是 11（如果参数值超过 11，则按 11 进行处理）。从 Oracle 11.2.0.2 开始，参数的最大值可设置为 1024。
- ASM_DISKSTRING：该参数非常重要，用来指定 Oracle ASM 实例可以发现的磁盘，多个路径之间以逗号分隔。例如，“/dev/asmdisk*”表示 ASM 实例只能使用/dev/目录下以 asmdisk 开头的磁盘。这里也可以使用“?”通配符，表示一个字符。

一般在 RAC 集群的规划阶段就已经进行了路径的规划，路径配置好之后可以查询 v$asm_disk 视图来确认这些磁盘。只有在 ASM_DISKSTRING 初始化参数中定义过的磁盘才可以被 ASM 使用。通过 V$ASM_DISK 视图中的参数 HEADER_STATUS 可以查看还没有被使用的磁盘和已使用的磁盘，代码如下：

```
SQL> select name, header_status, path from v$asm_disk;
NAME                         HEADER_STATU PATH
---------------------------------------- --------------------
                             MEMBER       /dev/asmdiskd
                             MEMBER       /dev/asmdiskb
                             CANDIDATE     /dev/asmdiskh
                             CANDIDATE     /dev/asmdiskf
                             MEMBER       /dev/asmdiske
                             MEMBER       /dev/asmdiskc
                             CANDIDATE     /dev/asmdiskg
DATA3                        MEMBER       AFD:DATA3
OCR_MGMT1                    MEMBER       AFD:OCR_MGMT1
DATA2                        MEMBER       AFD:DATA2
DATA1                        MEMBER       AFD:DATA1

11 rows selected.
```

参数 HEADER_STATUS 的状态有 FORMER、CANDIDATE、PROVISIONED 和 MEMBER，具体说明如下。

- FORMER：该磁盘之前属于一个磁盘组，后来从该磁盘组中删除了。
- CANDIDATE：该磁盘从来没有被使用过。
- PROVISIONED：根据不同的操作系统，使用相应的工具绑定磁盘。例如，Linux 操作系统使用 ASMLIB 和 ASMFD 工具，Windows 操作系统使用 ASMTOOL 和 ASMTOOLG 工具。
- MEMBER：现在属于某个磁盘组，无法被其他磁盘组使用。

这里需要注意以下几点。

- Oracle ASM 最多只能发现 10 000 个磁盘。如果 ASM_DISKSTRING 参数配置了超过 10 000 个磁盘，那么 ASM 只能发现前 10 000 个磁盘，其他磁盘将无法被使用。
- 只有参数 HEADER_STATUS 的状态是 CANDIDATE、PROVISIONED 和 FORMER 时，才可以在不指定 FORCE 参数的情况下将磁盘添加到磁盘组中。

- 当参数 HEADER_STATUS 的状态是 MEMBER 时，如果其所属的磁盘组不是 MOUNT（挂载）状态，那么可以通过 FORCE 参数强制将其添加到其他磁盘组中。

9.2 磁盘组管理

9.2.1 磁盘组属性

磁盘组属性是磁盘组的相关参数，有些参数属性可以在磁盘组创建或修改时设置，而有些参数属性只能在创建时设置。

可以通过 V$ASM_DISKGROUP 视图或 ASMCMD 的 lsattr 命令查看磁盘组的属性。可以通过 ALTER DISKGROUP、CREATE DISKGROUP 和 ASMCMD 的 setattr、mkdg 命令设置磁盘组属性。

通过 V$ASM_DISKGROUP 视图查看磁盘组的属性，代码如下：

```
    SELECT SUBSTR (dg.name, 1, 12) AS diskgroup,
     SUBSTR (a.name, 1, 24) AS name,
     SUBSTR (a.VALUE, 1, 24) AS VALUE,
     read_only
 FROM V$ASM_DISKGROUP dg, V$ASM_ATTRIBUTE a
WHERE    dg.name = 'DATA'
     AND dg.group_number = a.group_number
     AND a.name NOT LIKE '%template%';
DISKGROUP    NAME           VALUE                 READ_ON
------------ ------------------------ ------------------------
-------
DATA    idp.type                  dynamic          N
DATA    vam_migration_done        true             Y
DATA    disk_repair_time          12.0h            N
DATA    phys_meta_replicated      true             Y
DATA    failgroup_repair_time    24.0h             N
DATA    thin_provisioned         FALSE             N
DATA    preferred_read.enabled   FALSE             N
DATA    ate_conversion_done      true              Y
DATA    sector_size              512               N
DATA    logical_sector_size      512               N
DATA    content.type             data              N
```

使用 lsattr 命令查看磁盘组的属性，代码如下：

```
ASMCMD> lsattr -lm -G data
Group_Name  Name                          Value          RO  Sys
DATA        access_control.enabled        FALSE          N   Y
DATA        access_control.umask           066           N   Y
DATA        appliance._partnering_type    GENERIC        Y   Y
DATA        ate_conversion_done           true           Y   Y
DATA        au_size                       4194304        Y   Y
DATA        cell.smart_scan_capable       FALSE          N   N
DATA        cell.sparse_dg                allnonsparse   N   N
DATA        compatible.advm               18.0.0.0.0     N   Y
DATA        compatible.asm                18.0.0.0.0     N   Y
DATA        compatible.rdbms              10.1.0.0.0     N   Y
DATA        content.check                 FALSE          N   Y
DATA        content.type                  data           N   Y
DATA        content_hardcheck.enabled     FALSE          N   Y
DATA        disk_repair_time              12.0h          N   Y
DATA        failgroup_repair_time         24.0h          N   Y

ASMCMD> lsattr -G data -l %compat*
Name              Value
compatible.advm   18.0.0.0.0
compatible.asm    18.0.0.0.0
compatible.rdbms  10.1.0.0.0
```

修改 DATA 磁盘组的相关属性，代码如下：

```
ASMCMD> setattr -G data compatible.rdbms 18.0
ASMCMD> lsattr -G data -l %compat*
Name              Value
compatible.advm   18.0.0.0.0
compatible.asm    18.0.0.0.0
compatible.rdbms  18.0
```

通过前面的查询可以看到磁盘组的属性较多，下面是几个常见属性的说明。

- AU_SIZE：指定磁盘组的 AU 大小，只能在创建磁盘组时指定，指定后无法修改。
- COMPATIBLE.ASM：指定可以使用磁盘组的 Oracle ASM 实例的最低软件版本，该参数值还会影响磁盘上 Oracle ASM 元数据的数据结构的格式。
- COMPATIBLE.RDBMS：指定可以使用磁盘组的数据库实例的最低数据库版本。

- COMPATIBLE.ADVM：指定磁盘组是否可以包含 Oracle ADVM，该参数值必须设置为 11.2 或更高。在设置该参数之前，COMPATIBLE.ASM 参数值必须为 11.2 或更高。该属性的默认值取决于 Oracle ASM 的版本。
- DISK_REPAIR_TIME：指定修复磁盘到其在线(online)的时间间隔。该参数只能在磁盘组创建后修改，并且只能用于 Normal 和 High 冗余类型的磁盘组。
- FAILGROUP_REPAIR_TIME：用于指定修复出现问题的故障组的时间间隔，默认值是 24 小时（24h）。该参数只能用于 Normal 和 High 冗余类型的磁盘组。

9.2.2 创建磁盘组

创建磁盘组时需要指定以下几个属性：

- 磁盘组名称；
- 冗余类型；
- 所需磁盘；
- 指定故障组（可选）；
- 磁盘组属性（可选）。

创建磁盘组的方式有以下三种。

1. 使用 ASMCMD 工具（很少使用）

（1）创建 XML 文件，代码如下：

```
<dg name="data" redundancy="normal">
  <fg name="fg1">
    <dsk string="/dev/asmdiskb"/>
  </fg>
  <fg name="fg2">
    <dsk string="/dev/asmdiskc"/>
  </fg>
  <fg name="fg3">
    <dsk string="/dev/asmdiskd"/>
  </fg>
  <a name="compatible.asm" value="18.0"/>
  <a name="compatible.rdbms" value="18.0"/>
  <a name="compatible.advm" value="18.0"/>
```

```
</dg>
```

（2）通过 XML 文件创建磁盘组，代码如下：

```
ASMCMD [+] > mkdg data_config.xml
```

2. 使用 SQL 语句（常用）

这是最常用的一种创建磁盘组的方式，也是最方便的一种方式。下面创建一个冗余类型是 Normal 的 DATA 磁盘组，代码如下：

```
CREATE DISKGROUP data NORMAL REDUNDANCY
  FAILGROUP fg1 DISK
    '/dev/asmdiskb' NAME diskb,
  FAILGROUP fg2 DISK
    '/dev/asmdiskc' NAME diskc,
  FAILGROUP fg3 DISK
    '/dev/asmdiskd' NAME diskd,
  ATTRIBUTE 'au_size'='4M',
    'compatible.asm' = '18.0',
    'compatible.rdbms' = '18.0',
    'compatible.advm' = '18.0';
```

3. 使用 ASMCA 工具（常用）

这种方式直接调用 ASMCA 工具在图形界面上进行操作，比较方便，但要求图形界面的支持，如图 9-1 所示，而且有时远程调用图形不太方便。

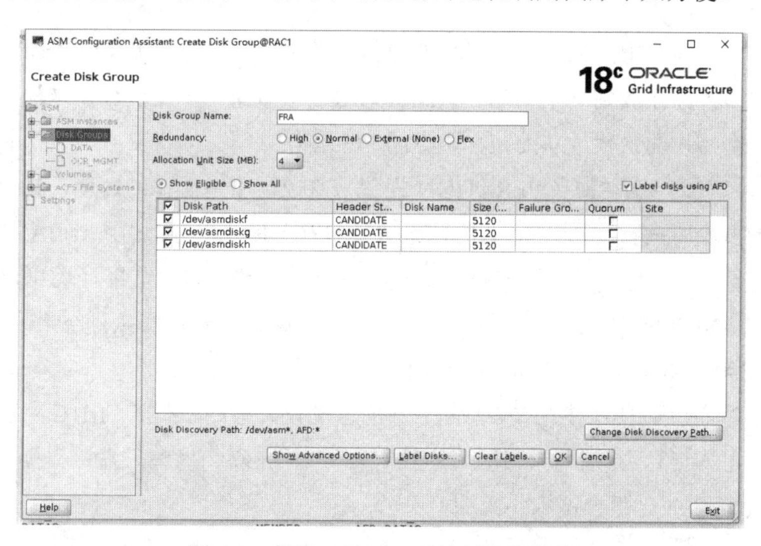

图 9-1　使用 ASMCA 工具创建磁盘组

9.2.3 删除磁盘组

删除磁盘组可以使用 ASMCA、ASMCMD 和 SQLPLUS 这几种方式。在删除时可以指定 INCLUDING CONTENTS 来删除磁盘组中包含的所有文件，默认是 EXCLUDING CONTENTS。

在删除磁盘组时，ASM 实例必须处于启动状态，对应的磁盘组必须是 MOUNT 状态并且磁盘组中的文件没有被使用。在执行 DROP DISKGROUP 命令时，磁盘组完全被删除后才会返回结果。

如果 ASM 实例使用的是 SPFILE 参数文件，那么 Oracle ASM 在删除磁盘组的同时也会将磁盘组从 ASM_DISKGROUPS 初始化参数中删除。如果 ASM 实例使用的是 PFILE 参数文件，则需要在下一次关闭和重新启动 Oracle ASM 实例之前，手动从 ASM_DISKGROUPS 初始化参数中删除磁盘组的名称。

如果磁盘组出现故障，无法为 MOUNT 状态，则必须使用 FORCE 参数来强行删除磁盘组。注意，一定要确认磁盘组没有被使用，因为添加 FORCE 参数后，Oracle 就不再去验证磁盘组是否正在被使用了。

在 SQLPLUS 中，删除磁盘组的语法如下：

```
SQL>DROP DISKGROUP data [FORCE];
```

9.2.4 磁盘组的再平衡

在生产环境中，磁盘组最常见的操作就是添加、删除、调整。Oracle 建议尽量将多个操作放到一个 ALTER DISKGROUP 命令中一次执行，这样可以减少再平衡操作的次数，从而减小磁盘的 I/O 压力。

可以查询 V$ASM_OPERATION 视图来监控再平衡操作的速度，预估需要完成的剩余时间。也可以在 ALTER DISKGROUP 语句中使用 REBALANCE WAIT 选项等待再平衡操作完成后返回 SQL 语句的执行结果，还可以使用 REBALANCE NOWAIT 选项不等再平衡操作完成就立刻返回执行结果。一般选择 REBALANCE NOWAIT 选项，在后台进行再平衡操作。

当然，使用 REBALANCE WAIT 选项也没有关系，直接按 Crtl+C 组合键，命令会进入后台继续执行，这也就意味着添加、删除、调整磁盘这类操作是无法通过按 Crtl+C 组合键来终止的。

1. 添加磁盘

在添加磁盘之前，要确认 V$ASM_DISK 中已经有该磁盘（通过参数 ASM_DISKSTRING 控制）。

（1）查看可用磁盘

通过查询视图和参数能够确认可用的磁盘，代码如下。如果添加了错误的磁盘，则可能会造成其他磁盘组的异常。

```
SQL> show parameter asm_diskstring

NAME                                 TYPE        VALUE
------------------------------------ ----------- ------------------------------
asm_diskstring                       string      /dev/asm*


SQL> set lin 150
SQL> col path for a30
SQL> select name,HEADER_STATUS,path from v$asm_disk order by 1;
NAME                           HEADER_STATU PATH
------------------------------ ------------
------------------------------
DATA1                          MEMBER       AFD:DATA1
DATA2                          MEMBER       AFD:DATA2
DATA3                          MEMBER       AFD:DATA3
OCR_MGMT1                      MEMBER       AFD:OCR_MGMT1
                               MEMBER       /dev/asmdiske
                               CANDIDATE    /dev/asmdiskg
                               CANDIDATE    /dev/asmdiskf
                               CANDIDATE    /dev/asmdiskh
                               MEMBER       /dev/asmdiskb
                               MEMBER       /dev/asmdiskd
                               MEMBER       /dev/asmdiskc

rows selected.
```

（2）查看磁盘组的相关信息

在添加磁盘之前要确认相关磁盘组中故障组的信息，默认情况下是一个磁盘作为一个故障组，但是实际生产环境中是一个存储节点作为一个故障组，所以要根据之前磁盘组的配置来进行相关的操作，代码如下：

```
SQL> SELECT SUBSTR(dg.name,1,16) AS diskgroup, SUBSTR(d.name,1,16)
```

```
AS asmdisk,
       d.mount_status, d.state, SUBSTR (d.failgroup,1,16) AS
failgroup
       FROM V$ASM_DISKGROUP dg, V$ASM_DISK d WHERE dg.group_number
= d.group_number;
   DISKGROUP      ASMDISK        MOUNT_S STATE   FAILGROUP
   ---------------- ---------------- ------- --------
----------------
   DATA           DATA3          CACHED  NORMAL   FG3
   DATA           DATA2          CACHED  NORMAL   FG2
   DATA           DATA1          CACHED  NORMAL   FG1
   OCR_MGMT       OCR_MGMT1      CACHED  NORMAL   OCR_MGMT1
```

这里可以看到，data 磁盘组是一个磁盘作为一个故障组，所以在添加磁盘时可以不用指定故障组，因为 Oracle 会自动将该磁盘作为一个故障组并自动命名。

（3）添加磁盘

这里新添加的磁盘是从来没有使用过的，如果要添加的磁盘之前是某个磁盘组中的磁盘，那么可以通过 FORCE 参数将其添加到另一个磁盘组中，但是要仔细确认该磁盘在原磁盘组中已经不再使用了。添加磁盘的代码如下：

```
SQL>alter diskgroup data add disk '/dev/asmdiskh' rebalance power
5 nowait [force];
```

假如要向磁盘组 data 的 D3SSD 故障组中添加一个磁盘，那么代码如下：

```
SQL>alter diskgroup data add failgroup d3ssd disk '/dev/asmdiskh'
rebalance power 5 nowait [force];
```

（4）查看被添加磁盘的再平衡预计完成时间

在添加磁盘后，Oracle 会自动将磁盘组中的磁盘数据复制到新磁盘中。可以通过 v$asm_operation 视图查看添加磁盘后数据再平衡的操作时间，代码如下：

```
SQL>select group_number,operation,state,power,est_work,est_
rate,est_minutes from v$asm_operation;
```

2．删除磁盘

可以使用 DROP DISKS IN FAILGROUP 子句删除指定故障组中的磁盘。在删除磁盘时，会将所有文件从已删除的磁盘移动到磁盘组的其他磁盘中，也就是进行再平衡。如果其他磁盘上没有足够的可用空间，则删除磁盘操作可能会失败。所以，最好在删除磁盘的同时执行添加磁盘操作，就可以确保有足够的空间来进行再平衡操作。

删除磁盘 diskh，如果磁盘 diskh 已经无法访问，那么需要添加 FORCE 参数，代码如下：

```
SQL>alter diskgroup data drop disk diskh [force];
```

如果删除磁盘后想添加一个磁盘，那么尽量在一个操作中完成，代码如下：

```
SQL>alter diskgroup data drop disk diskh add failgroup fg1 disk
'/dev/asmdiski' name diski;
```

这里需要注意以下两点。

- 使用 DROP FORCE 命令删除磁盘，可能会降低磁盘组的冗余级别。如果再平衡数据期间其他磁盘再出现问题，那么就会造成数据丢失。
- 删除后的磁盘如果想继续使用，需要查看 V$ASM_DISK 视图中的 HEADER_STATUS 列值是否是 FORMER。

3．替换磁盘

替换磁盘和之前讲解的删除磁盘、添加磁盘类似，但替换磁盘的操作并不是通过删除磁盘再添加磁盘来实现的。直接替换磁盘的操作比通过删除磁盘再添加磁盘的操作具有更高的效率。

以下代码将磁盘组 data 中的磁盘 diskh 替换为/dev/asmdiski，磁盘组中的其他配置都不发生变化，包括故障组、磁盘名称等。

```
SQL>ALTER DISKGROUP data REPLACE DISK diskh WITH '/dev/ asmdiski
' POWER 3;
```

4．重命名磁盘

在重命名磁盘时，磁盘所在的磁盘组要以受限的模式挂载，并且磁盘组中的所有磁盘必须是在线状态，否则操作会失败。重命名磁盘的代码如下：

```
SQL> alter diskgroup fra mount restricted;
SQL> alter diskgroup fra rename disk 'fra1_0001' to 'fra2_0001',
'fra1_0002' to 'fra2_0002';
```

需要注意，如果磁盘被其他方式所绑定，例如 ASMLIB、ASMFD、EXADATA 和 ASMTOOL，并且这些标签名成为磁盘的名称，则在这种情况下执行 ALTER

DISKGROUP RENAME DISKS SQL 语句无效，磁盘不会被重命名。

5．手动再平衡

从实际运维经验来看，对 ASM 磁盘组的操作非常少，如果需要添加或删除磁盘，那么可能需要手动指定再平衡的级别。

默认情况下执行 ALTER DISKGROUP…REBALANCE 命令会立刻返回结果，再平衡操作会在后台执行，当然也可以指定 WAIT 参数等待再平衡操作完成。同样的，无法通过按 Ctrl+C 组合键来终止再平衡操作。

如下代码对磁盘组 data 进行手动再平衡操作，POWER 值为 5，直到再平衡操作完成才返回结果。

```
SQL>ALTER DISKGROUP data REBALANCE RESTORE POWER 5 WAIT;
```

再平衡操作有以下几个注意事项。

- 在再平衡操作期间，如果因为更改存储配置或停电等导致再平衡操作终止，那么 Oracle 会重新启动再平衡操作。此外，如果再平衡操作因用户操作错误而失败，则可能需要手动再平衡。
- 在 Oracle RAC 环境中，ALTER DISKGROUP ... REBALANCE 语句只会在单个节点上运行。
- Oracle ASM 可以在指定的实例上一次只执行一个磁盘组再平衡。如果在单个节点上的不同磁盘组上启动了多个再平衡操作，那么 Oracle 会在其他节点上并行处理这些操作（如果有其他节点可用的话）；否则，再平衡操作在单个节点上串行执行。
- 在再平衡操作期间，如果 Oracle ASM 实例挂掉，则再平衡操作仍会继续。
- 再平衡子句（及其关联的 POWER 和 WAIT / NOWAIT 关键字）也可用于添加、删除和调整磁盘大小的 ALTER DISKGROUP 命令。

6．清理磁盘组

Oracle ASM 磁盘的清理通过搜索很少被读取的数据来提高可用性和可靠性。清理磁盘时会检查逻辑数据损坏情况并在 Normal 和 High 冗余磁盘组中进行自动修复。清理过程使用镜像副本数据来修复逻辑损坏。磁盘清理可与磁盘组再平衡操作一起执行，从而减少 I/O。磁盘清理对 I/O 的影响很小。

清理磁盘组 data 的代码如下：

```
SQL> ALTER DISKGROUP data SCRUB POWER LOW;
```

清理数据文件的代码如下：

```
SQL> ALTER DISKGROUP data SCRUB FILE '+DATA/ORCL/DATAFILE/
example.266.806582193' REPAIR POWER HIGH FORCE;
```

清理磁盘的代码如下：

```
SQL> ALTER DISKGROUP data SCRUB DISK DATA_0005 REPAIR POWER HIGH
FORCE;
```

9.2.5　管理磁盘组容量

在创建冗余类型是 Normal 或 High 的磁盘组时，每个磁盘组必须有足够的容量来存放在一个或两个磁盘组发生故障后数据的再平衡操作。在一个或多个磁盘发生故障后，在恢复所有数据冗余的过程中，需要用磁盘组中剩余磁盘的空间来存放出现故障的磁盘中的数据。如果剩余空间不足，则某些文件可能会降低冗余级别。冗余级别的降低意味着文件中的一个或多个扩展区没有按照预期的冗余级别进行镜像存储。

所以，在规划磁盘组容量时需要考虑以下两点。

- Normal 冗余磁盘组：磁盘组中最好有足够的可用空间，以允许丢失一个故障组中的所有磁盘。可用空间大小应等于最大故障组的大小。
- High 冗余磁盘组：最好有足够的可用空间来应对两个故障组中所有磁盘的丢失。可用空间大小应等于两个最大故障组的大小总和。

1. V$ASM_DISKGROUP 视图中相关字段的解释说明

如果磁盘组的类型为 Flex 或 Extend，则 V$ASM_DISKGROUP 视图的 REQUIRED_MIRROR_FREE_MB 列和 USABLE_FILE_MB 列中的值为 0。在这种情况下，不是表示 REQUIRED_MIRROR_FREE_MB 列和 USABLE_FILE_MB 列的列值为 0。

- REQUIRED_MIRROR_FREE_MB：磁盘组发生故障后磁盘组中可用的空间量，这里的空间是在不添加新存储的情况下恢复冗余的空间。该空间可确保有足够的故障组来恢复冗余。另外，这里的故障指磁盘的永久性损坏，而不是磁盘离线（offline）后重新在线（online）的情况。

此列中显示的空间大小会考虑镜像数的影响，该值的计算方法如下。

① 有两个以上 REGULAR 故障组的正常冗余磁盘组，该值是最大故障组中所有磁盘的空间总和。

② 有三个以上 REGULAR 故障组的高冗余磁盘组，该值是前两个最大故障组中所有磁盘的空间总和。

- USABLE_FILE_MB：表示在满足完全冗余的情况下，可用的空间大小。该值的计算方法是从磁盘组的总可用空间中减去 REQUIRED_MIRROR_FREE_MB 的值，然后除以镜像数，如下所示：

```
(FREE_MB - REQUIRED_MIRROR_FREE_MB) / 2 = USABLE_FILE_MB
```

- TOTAL_MB：是磁盘组的总可用容量，以 B 为单位。该值会将磁盘头占用的空间去掉。磁盘头空间取决于 Oracle ASM 磁盘和 Oracle ASM 文件的数量，通常约为总原始存储容量的 1%。例如，磁盘组中的所有磁盘空间加起来是 100GB，那么 TOTAL_MB 大概为 99GB。
- FREE_MB：是磁盘组未使用的空间。在某些情况下，FREE_MB 显示是否有使用的空间。有时其中一个磁盘已满（可能是每个磁盘总空间大小不同），会因磁盘组不平衡而导致数据写入失败。

2．USABLE_FILE_MB 出现负值的原因

由于 FREE_MB、REQUIRED_MIRROR_FREE_MB 和 USABLE_FILE_MB 之间存在关系，所以 USABLE_FILE_MB 可能变为负数。看到是负值也不要紧张，需要注意以下几点。

- USABLE_FILE_MB 的值取决于 FREE_MB 值的大小，为负值时可能无法创建新文件。
- 如果某个磁盘出现故障，则可能导致某些文件降低冗余级别。

所以，如果 USABLE_FILE_MB 是负值，那么强烈推荐添加更多的磁盘以增加足够的容量。

9.2.6 磁盘组的性能和可伸缩性

磁盘组的性能不仅受硬件的限制，而且有时受磁盘组的相关配置的影响也非常大。

1. 磁盘组的数量

- 一个磁盘组中的磁盘应该大小相同且硬件性能相似（最好相同）。如果在大小和性能方面有不同类型的磁盘，则可以分开创建多个磁盘组。
- 不同存储功能对应不同的磁盘组，比如为数据文件创建 DATA 磁盘组、为备份文件创建 FRA 磁盘组、为 OCR 和 VOTING DISK 创建 OCRVD 磁盘组。

2. 将磁盘分组

分组时尽量将大小相同、性能相似的磁盘放到一个磁盘组中，这样可以平衡每个磁盘的负载和使用空间。

3. ASM 的存储限制

Oracle ASM 可以提供近乎无限大的存储容量，但有些限制还需要注意一下。对磁盘组、磁盘、文件数量的限制如下。

- Oracle 12c R1 及以后的版本中最多支持 511 个磁盘组。
- 每个磁盘组最多支持 10 000 个磁盘。
- 整个 ASM 最多支持 65 530 个磁盘。
- 每个磁盘组最多支持存储 100 万个文件。

如果 COMPATIBLE.ASM 或 COMPATIBLE.RDBMS 参数值小于 12.1，则 Oracle ASM 会有以下限制。

- 每个磁盘最大支持 2TB。
- 整个 ASM 存储最大支持 20PB。

如果 COMPATIBLE.ASM 或 COMPATIBLE.RDBMS 参数值大于 12.1，则 Oracle ASM 有以下限制。

- AU=1MB 时，单个磁盘最大支持 4PB。
- AU=2MB 时，单个磁盘最大支持 8PB。
- AU=4MB 时，单个磁盘最大支持 16PB。
- AU=8MB 时，单个磁盘最大支持 32PB。
- 整个 ASM 存储最大支持 320EB。

综上可知，磁盘组最大支持的大小等于最大支持的磁盘数乘以每个磁盘最大支持的大小。另外，磁盘组最大支持的大小还受操作系统的版本和 DB_BLOCK_SIZE 值的影响。Oracle 现在支持的最大容量可以说几乎没有限制，正常情况下绝对够用。

9.2.7 磁盘组的兼容性

磁盘组的兼容性参数的作用是使用对应版本中的新特性，并且参数只能从低往高设置，而不能从高往低设置。例如，将磁盘组的兼容性参数设置为 18.0，那么就可以使用 Oracle 18c 中关于磁盘组的新特性了。

磁盘组的兼容性参数有 COMPATIBLE.ASM、COMPATIBLE.RDBMS 和 COMPATIBLE.ADVM。参数 COMPATIBLE.ASM 和 COMPATIBLE.RDBMS 分别用于为 Oracle ASM 和数据库实例指定使用的软件最低版本号。COMPATIBLE.ADVM 参数用于确定 Oracle ASM Dynamic Volume Manager 功能是否可以在磁盘组中创建卷。

可以通过 CREATE DISKGROUP、ALTER DISKGROUP 和 ASMCMD 中的 setattr 命令来设置、修改磁盘组的兼容性。注意，COMPATIBLE.ASM 参数的值一定要比磁盘组其他相关属性参数的值大。

设置磁盘组的兼容性，代码如下：

```
CREATE DISKGROUP data NORMAL REDUNDANCY
  FAILGROUP fg1 DISK
    '/dev/asmdiskb' NAME diskb,
  FAILGROUP fg2 DISK
    '/dev/asmdiskc' NAME diskc,
  FAILGROUP fg3 DISK
    '/dev/asmdiskd' NAME diskd,
  ATTRIBUTE 'au_size'='4M',
    'compatible.asm' = '18.0',
    'compatible.rdbms' = '18.0',
    'compatible.advm' = '18.0';
```

修改磁盘组的兼容性，代码如下：

```
ALTER DISKGROUP data SET ATTRIBUTE 'compatible.asm' = '18.0';
ALTER DISKGROUP data SET ATTRIBUTE 'compatible.rdbms' = '18.0',
ALTER DISKGROUP data SET ATTRIBUTE 'compatible.advm' = '18.0';
```

如表 9-1 所示是 ASM 相关新特性对应的兼容性参数值。

表 9-1　ASM 相关新特性对应的兼容性参数值

磁盘组的特性	COMPATIBLE.ASM	COMPATIBLE.RDBMS	COMPATIBLE.ADVM
支持更大的 AU（32MB 或 64 MB）	≥ 11.1	≥11.1	n/a
可通过 V$ASM_ATTRIBUTE 视图查看属性	≥11.1	n/a	n/a
快速镜像同步	≥11.1	≥11.1	n/a
可变的扩展	≥11.1	≥11.1	n/a
Exadata 存储	≥11.1.0.7	≥11.1.0.7	n/a
支持 OCR 和 VOTING 文件存储到磁盘组中	≥11.2	n/a	n/a
修改 Sector 的大小	≥11.2	≥11.2	n/a
Oracle ASM SPFILE 可存放到磁盘组中	≥11.2	n/a	n/a
Oracle ASM File Access Control	≥11.2	≥11.2	n/a
ASM_POWER_LIMIT（最大值可设置为 1024）	≥11.2.0.2	n/a	n/a
磁盘组的复制状态	≥12.1	n/a	n/a
管理磁盘组中的共享密码文件	≥12.1	n/a	n/a
可以支持单个磁盘大小超过 2TP	≥12.1	≥12.1	n/a
支持重新同步检查点	≥12.1.0.2	≥12.1.0.2	n/a
支持 LOGICAL_SECTOR_SIZE	≥12.2	n/a	n/a
支持 Flex 和 ExtendED 类型的磁盘组	≥12.2	≥12.2	n/a
支持 SCRUB_ASYNC_LIMIT	≥12.2	n/a	n/a
支持 PREFERRED_READ.ENABLED	≥12.2	n/a	n/a
支持非限制模式下降 NORMAL 或 HIGH 冗余类似的磁盘组转换为 Flex 类型磁盘组	≥18.0	≥12.2	n/a
Flex 类型磁盘组支持多租户复制	≥18.0	≥18.0	n/a

如表 9-2 所示是 AU_SIZE 为 1MB 时磁盘组支持的最大文件大小。

表 9-2 AU_SIZ 为 1MB 时磁盘组支持的最大文件大小

冗 余 类 型	COMPATIBLE.RDBMS = 10.1	COMPATIBLE.RDBMS ≥11.1
External	16 TB	128 TB
Normal	5.8 TB	93 TB
High	3.9 TB	62 TB

如表 9-3 所示是 AU_SIZE 为 4MB 时磁盘组支持的最大文件大小。

表 9-3 AU_SIZE 为 4MB 时磁盘组支持的最大文件大小

冗 余 类 型	COMPATIBLE.RDBMS = 10.1	COMPATIBLE.RDBMS≥11.1
External	64 TB	128 TB
Normal	32 TB	128 TB
High	21 TB	128 TB

9.3 查看 ASM 信息

对于已有的磁盘组及其他 ASM 信息，可以通过相关视图来查看，这些视图如表 9-4 所示。

表 9-4 查看 ASM 信息的视图

视 图	说 明
V$ASM_ALIAS	显示已挂载的磁盘组中存在的别名相关信息
V$ASM_ATTRIBUTE	显示磁盘组相关的属性。除可以用 CREATE DISKGROUP 和 ALTER DISKGROUP 语句指定属性外，视图还显示自动创建的其他属性。仅显示 COMPATIBLE.ASM 被设置为 11.1 或更高的磁盘组属性
V$ASM_AUDIT_CLEAN_EVENTS	显示有关审计跟踪清除或清除事件历史记录的信息
V$ASM_AUDIT_CLEANUP_JOBS	显示有关已配置的审计跟踪清除的信息
V$ASM_AUDIT_CONFIG_PARAMS	显示有关当前配置的审计跟踪属性的信息
V$ASM_AUDIT_LAST_ARCH_TS	显示有关为审计跟踪清除或清除设置的上次存档时间戳的信息
V$ASM_CLIENT	在 Oracle ASM 实例中，使用 Oracle ASM 实例管理的磁盘组标识数据库。在 Oracle 数据库实例中，如果数据库打开了 Oracle ASM 文件，则会显示有关 Oracle ASM 实例的信息
V$ASM_DBCLONE_INFO	在 Oracle ASM 实例中，显示源数据库、复制数据库及其文件组之间的关系

续表

视　图	说　明
V$ASM_DISK	显示 Oracle ASM 实例发现的每个磁盘的信息，包括还没有使用的磁盘。每次查询该视图时，Oracle 都会重新查找磁盘的动作。由于查找磁盘非常耗费资源，因此不建议将此视图用于监控脚本
V$ASM_DISK_IOSTAT	显示每个 Oracle ASM 客户端的磁盘 I/O 统计信息。在 Oracle 数据库实例中，仅显示该实例的信息
V$ASM_DISK_STAT	与 V$ASM_DISK 视图具有相同的列，但是为了减少性能开销，在查询时不再查找磁盘。V$ASM_DISK_STAT 视图仅显示已挂载的磁盘组中的磁盘信息。建议使用 V$ASM_DISK_STAT 视图来监控脚本。要显示所有磁盘的信息，需要查看 V$ASM_DISK 视图
V$ASM_DISKGROUP	显示磁盘组（数量、名称、大小、状态和冗余类型）的相关信息。和 V$ASM_DISK 视图一样，每次查询都会自动执行查找磁盘操作，消耗资源
V$ASM_DISKGROUP_STAT	与 V$ASM_DISKGROUP 视图具有相同的列，但是在查询时不会查找磁盘
V$ASM_ESTIMATE	显示 Oracle ASM 磁盘组再平衡和重新同步的操作时间、工作量的预估信息
V$ASM_FILE	显示已挂载的磁盘组中的 ASM 文件信息
V$ASM_FILEGROUP	显示已挂载的磁盘组中的 ASM 文件组信息
V$ASM_FILEGROUP_FILE	显示与已挂载的磁盘组中的 ASM 文件组相关的文件信息
V$ASM_FILEGROUP_PROPERTY	显示已挂载的磁盘组中的文件组或每个文件组的每种文件类型的信息
V$ASM_OPERATION	在 Oracle ASM 实例中，显示 Oracle ASM 实例中的长时间运行操作。在 Oracle 数据库实例中，不显示任何数据
V$ASM_QUOTAGROUP	显示 Oracle ASM 配额组的相关信息
V$ASM_TEMPLATE	显示已挂载的磁盘组中的模板信息
V$ASM_USER	显示已连接数据库实例的有效操作系统用户名和文件所有者的名称
V$ASM_USERGROUP	包含每个 Oracle ASM 文件访问控制组的创建者
V$ASM_USERGROUP_MEMBER	包含每个 Oracle ASM 文件访问控制组的成员

用户需要注意以下两点。

- 以上视图中的 GROUP_NUMBER 表示磁盘组编号，磁盘组编号不是固定不变的。每次重新挂载磁盘组时，磁盘组编号都可能发生变化。
- V$ASM_FILE 视图中的 REDUNDANCY_LOWERED 列不再显示冗余级别降低的相关文件信息，该列已经废弃并且列的值一直是 U。

第 10 章

ASM Filter Driver （ASMFD）

Oracle 从 12.1.0.2 版本开始支持 ASMFD（仅限 Linux 平台），也支持 Solaris 平台。ASMFD 是 ASMLIB 和 UDEV 的替代产品，实际上 ASMFD 也用到了 UDEV。更为关键的是，ASMFD 支持验证写 I/O 的请求，如果是非法的写操作，则 Oracle 会拒绝，从而更好地保护磁盘数据不被删除或覆盖。

10.1 ASMFD 的概念

ASMFD 是 Oracle ASM 磁盘 I/O 路径中的内核模块。Oracle ASM 使用 Filter Driver 来验证对 Oracle ASM 磁盘的写入 I/O 请求是否合法、是否是数据库相关的正常请求，如果不是，Oracle 会拒绝该操作对磁盘进行写入。

从 Oracle 12.2 开始，如果操作系统上已经安装了 ASMLIB，那么将无法使用 ASMFD，只能二选一。

ASMFD 的主要特性如下。

- 拒绝非 Oracle 的 I/O 操作：在出现 ASMFD 之前，所有用户都可以对磁盘内容进行读写，不管是否合法，而误操作会导致磁盘数据被覆盖，造成数据无法恢复。ASMFD 只允许特定的 Oracle 接口进行写操作，并拒绝所有非 Oracle 应用程序对磁盘进行写操作，这样就大大降低了磁盘数据被损坏的概率。
- 减少操作系统的资源使用：ASM 实例中包含了大量的进程或线程，在没有使用 ASMFD 的情况下，进程要对某个磁盘进行写操作，就要取得一个文件描述符（File Descriptor）。一旦有数千个进程，就需要数千个文件描述符，这样就会消耗更多的系统资源。而 ASMFD 提供了一个统一的接口，所有进程都可以通过该接口对磁盘进行操作，这样就减少了文件描述符，从而减少了资源的使用。

- 永久绑定磁盘：ASMFD 绑定过的磁盘，系统重启后无须重新进行绑定。
- 更快的节点恢复：当集群同步服务（CSS）不能正常工作时，init.d 脚本就会通知节点。使用这种机制，故障节点将被隔离出来，以确保集群中的其他节点正常运行。这种解决方案虽然有效，但是代价很高，因为需要重新启动节点并重新启动所有需要的进程。ASMFD 允许集群在不重新引导的情况下执行节点级隔离。因此，使用 ASMFD 可以通过重新启动软件栈（Software stack）而不是重新启动整个节点来实现相同的目标。

ASMFD 在集群环境中的逻辑位置如图 10-1 所示。

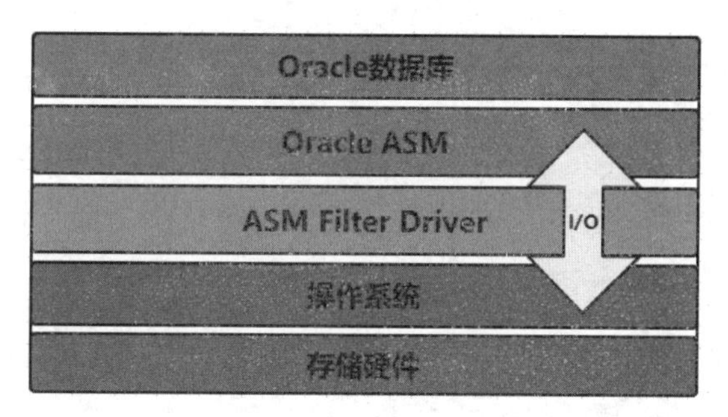

图 10-1　ASMFD 在集群环境中的逻辑位置

10.2　配置 ASMFD

可以在安装 Grid 时配置 ASMFD，也可以在安装 Grid 后再进行配置。

10.2.1　在安装 Grid 时配置 ASMFD

在安装 Grid 时勾选“Configure Oracle ASM Filter Driver”复选框即可，如图 10-2 所示。

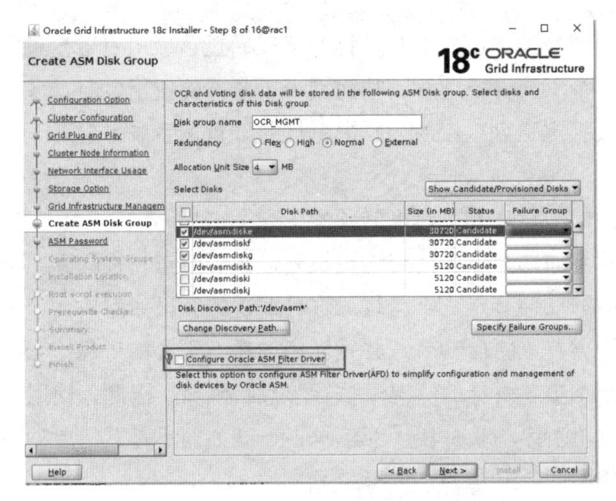

图 10-2　在安装 Grid 时配置 ASMFD

在创建磁盘组时，可以通过 ASMCA 来配置 ASMFD，如图 10-3 所示。

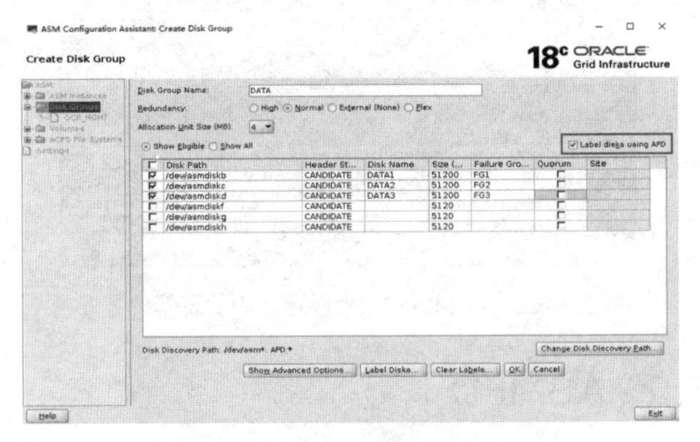

图 10-3　在创建磁盘组时配置 ASMFD

10.2.2　在安装 Grid 后配置 ASMFD

下面是笔者搭建的测试环境，在安装 Grid 时没有配置 ASMFD，所以这里并没有对应的内核模块。

```
[grid@18cASM software]$  lsmod|grep ora
oracleacfs          5438460  0
oracleadvm          1100111  0
oracleoks            732987  2 oracleacfs,oracleadvm
```

下面进行相关的配置操作。

1. 更新 Disk String

（1）查看当前 Disk String，代码如下：

```
[grid@18cASM software]$ asmcmd dsget
parameter:/dev/sd*, /dev/asm*
profile:/dev/sd*,/dev/asm*
```

（2）更新 Disk String，代码如下：

```
[grid@18cASM software]$ asmcmd dsset '/dev/asm*,AFD:*'
```

（3）验证 Disk String，代码如下：

```
[grid@18cASM software]$ asmcmd dsget
parameter:/dev/asm*, AFD:*
profile:/dev/asm*,AFD:*
```

2. 启用 ASMFD

下面的操作需要在所有节点执行。在 RAC 环境中也需要在所有节点执行，可以通过 Rolling 方式来配置。

下面以节点 1 为例，其他节点的操作与其相同。

（1）用 root 账户关闭 HAS，如果有必要，可以加-f 参数强制关闭 HAS，代码如下：

```
[root@18cASM ~]# crsctl stop has
CRS-2791: Starting shutdown of Oracle High Availability
Services-managed resources on '18casm'
CRS-2673: Attempting to stop 'ora.OCRVD.dg' on '18casm'
CRS-2673: Attempting to stop 'ora.LISTENER.lsnr' on '18casm'
CRS-2677: Stop of 'ora.OCRVD.dg' on '18casm' succeeded
CRS-2673: Attempting to stop 'ora.asm' on '18casm'
CRS-2677: Stop of 'ora.LISTENER.lsnr' on '18casm' succeeded
CRS-2677: Stop of 'ora.asm' on '18casm' succeeded
CRS-2673: Attempting to stop 'ora.evmd' on '18casm'
CRS-2677: Stop of 'ora.evmd' on '18casm' succeeded
CRS-2673: Attempting to stop 'ora.cssd' on '18casm'
CRS-2677: Stop of 'ora.cssd' on '18casm' succeeded
CRS-2793: Shutdown of Oracle High Availability Services-managed
resources on '18casm' has completed
CRS-4133: Oracle High Availability Services has been stopped.
```

（2）检查操作系统版本是否支持 ASMFD，代码如下：

```
[grid@18cASM grid]$ cat /etc/redhat-release
Red Hat Enterprise Linux Server release 7.5 (Maipo)
[grid@18cASM grid]$ uname -a
Linux 18cASM 3.10.0-862.el7.x86_64 #1 SMP Wed Mar 21 18:14:51 EDT
2018 x86_64 x86_64 x86_64 GNU/Linux
[grid@18cASM grid]$
[grid@18cASM grid]$ acfsdriverstate -orahome $ORACLE_HOME
supported
ACFS-9200: Supported
```

这里要确保操作系统版本支持 ASMFD，在 Redhat 7.4 的系统中就可能报如下错误：

```
ACFS-9459: ADVM/ACFS is not supported on this OS version:
'3.10.0-693.el7.x86_64'
ACFS-9201: Not Supported
```

这个问题在 MOS 上的两篇文章中有记录，文章编号分别是 2303388.1 和 26247490.8。

（3）配置 ASMFD。

用 root 账户在/etc/profile 中配置 ORACLE_BASE 和 ORACLE_HOME，代码如下；否则会报错“ASMCMD-00003: ORACLE_BASE environment variable not set”。

```
[root@18cASM ~]# echo $ORACLE_BASE
/u01/app/grid
[root@18cASM ~]# echo $ORACLE_HOME
/u01/app/18.3.0/grid
```

配置 ASMFD，代码如下：

```
[root@18cASM ~]# asmcmd afd_configure
AFD-627: AFD distribution files found.
AFD-634: Removing previous AFD installation.
AFD-635: Previous AFD components successfully removed.
AFD-9294: updating file /etc/sysconfig/oracledrivers.conf
AFD-636: Installing requested AFD software.
AFD-637: Loading installed AFD drivers.
AFD-9321: Creating udev for AFD.
AFD-9323: Creating module dependencies - this may take some time.
AFD-9154: Loading 'oracleafd.ko' driver.
```

```
AFD-649: Verifying AFD devices.
AFD-9156: Detecting control device '/dev/oracleafd/admin'.
AFD-638: AFD installation correctness verified.
Modifying resource dependencies - this may take some time.
[root@18cASM ~]#
```

查看 ASMFD 的核心模块 oracleafd，代码如下：

```
[root@18cASM ~]# lsmod|grep ora
oracleafd            210307  0
oracleacfs          5438460  0
oracleadvm          1100111  0
oracleoks            732987  2 oracleacfs,oracleadvm
```

（4）验证 ASMFD 的状态，代码如下：

```
[root@18cASM ~]# asmcmd afd_state
ASMCMD-9526: The AFD state is 'LOADED' and filtering is 'ENABLED'
on host '18cASM'
[root@18cASM ~]#
```

（5）将所有磁盘纳入 ASMFD 管理。

在笔者的测试环境中，创建了很多 Disk，代码如下：

```
[root@18cASM ~]# fdisk -l|grep Disk
Disk /dev/sda: 42.9 GB, 42949672960 bytes, 83886080 sectors
Disk label type: dos
Disk identifier: 0x000e6d24
Disk /dev/sdb: 1073 MB, 1073741824 bytes, 2097152 sectors
Disk /dev/sdc: 1073 MB, 1073741824 bytes, 2097152 sectors
Disk /dev/sdd: 1073 MB, 1073741824 bytes, 2097152 sectors
Disk /dev/sde: 42.9 GB, 42949672960 bytes, 83886080 sectors
Disk /dev/sdf: 10.7 GB, 10737418240 bytes, 20971520 sectors
Disk /dev/sdg: 10.7 GB, 10737418240 bytes, 20971520 sectors
Disk /dev/sdh: 1073 MB, 1073741824 bytes, 2097152 sectors
Disk /dev/sdi: 1073 MB, 1073741824 bytes, 2097152 sectors
Disk /dev/sdj: 1073 MB, 1073741824 bytes, 2097152 sectors
Disk /dev/sdk: 1073 MB, 1073741824 bytes, 2097152 sectors
Disk /dev/sdl: 1073 MB, 1073741824 bytes, 2097152 sectors
Disk /dev/sdm: 1073 MB, 1073741824 bytes, 2097152 sectors
Disk /dev/sdn: 1073 MB, 1073741824 bytes, 2097152 sectors
```

对于没有使用过的磁盘，可以用 root 账户执行以下命令，让 ASMFD 来管理这些磁盘，注意这里使用的是--init，代码如下：

```
[root@18cASM ~]# asmcmd afd_label testdt1 /dev/sdm --init
ASMCMD-9521: AFD is already configured
[root@18cASM ~]# asmcmd afd_lslbl /dev/sdm
--------------------------------------------------------------
Label                    Duplicate  Path
===============================================================
TESTDT1                              /dev/sdm
```

对于已经使用过的磁盘，需要加--migrate 参数才能修改成功。当然，对于没有使用过的磁盘，在使用--init 没有成功的情况下，也可以使用--migrate，代码如下：

```
[root@18cASM ~]# asmcmd afd_label testdt2 /dev/sdn --migrate
[root@18cASM ~]# asmcmd afd_lslbl /dev/sdn
--------------------------------------------------------------
Label                    Duplicate  Path
==============================================================
TESTDT2                              /dev/sdn
```

另外要注意，之前 OCRVD 磁盘组使用的是传统 UDEV 方式配置的，代码如下：

```
SQL> select name,label,path from v$asm_disk;

NAME                 LABEL                 PATH
------------------------------ ------------------------------
OCRVD_0001                                 /dev/asmdiskc
OCRVD_0000                                 /dev/asmdiskb
OCRVD_0002                                 /dev/asmdiskd

[root@18cASM ~]# asmcmd afd_label ocr1 /dev/sdb --migrate
[root@18cASM ~]# asmcmd afd_label ocr2 /dev/sdc --migrate
[root@18cASM ~]# asmcmd afd_label ocr3 /dev/sdd --migrate
[root@18cASM ~]# asmcmd afd_lslbl
---------------------------------------------------------------
Label                    Duplicate  Path
===============================================================
OCR1                                /dev/sdb
OCR2                                /dev/sdc
OCR3                                /dev/sdd

[root@18cASM ~]# asmcmd afd_label testdt3 /dev/sdl --migrate
[root@18cASM ~]# asmcmd afd_label testdt4 /dev/sdk -migrate
[root@18cASM ~]# asmcmd afd_label testdt5 /dev/sdj --migrate
```

```
[root@18cASM ~]# asmcmd afd_label testdt6 /dev/sdi --migrate
[root@18cASM ~]# asmcmd afd_label testdt7 /dev/sdh -migrate
[root@18cASM ~]# asmcmd afd_label data1 /dev/sdg --migrate
[root@18cASM ~]# asmcmd afd_label data2 /dev/sdf --migrate
[root@18cASM ~]# asmcmd afd_label migr /dev/sde --migrate

[root@18cASM ~]# asmcmd afd_lsdsk
--------------------------------------------------------------
Label                    Filtering  Path
==============================================================
DATA1                     ENABLED   /dev/sdg
DATA2                     ENABLED   /dev/sdf
MIGR                      ENABLED   /dev/sde
OCR1                      ENABLED   /dev/sdb
OCR2                      ENABLED   /dev/sdc
OCR3                      ENABLED   /dev/sdd
TESTDT1                   ENABLED   /dev/sdm
TESTDT2                   ENABLED   /dev/sdn
TESTDT3                   ENABLED   /dev/sdl
TESTDT4                   ENABLED   /dev/sdk
TESTDT5                   ENABLED   /dev/sdj
TESTDT6                   ENABLED   /dev/sdi
TESTDT7                   ENABLED   /dev/sdh
```

如果是集群环境，则该步骤只需要在其中一个节点执行，在另一个节点执行 smcmd afd_scan 命令就可以扫描这些配置。

（6）启动 HAS，代码如下：

```
[root@18cASM ~]# crsctl start has
CRS-4123: Oracle High Availability Services has been started.
```

（7）将 afd_ds 设置为底层磁盘路径，以后就不需要手动配置 UDEV 文件了，代码如下：

```
[grid@18cASM grid]$ asmcmd afd_dsset '/dev/sd*'
[grid@18cASM grid]$ asmcmd afd_dsget
AFD discovery string: /dev/sd*
```

（8）移除原来的 UDEV 配置文件。由于使用了 ASMFD 方式来管理和配置磁盘，所以不再需要 UDEV 配置文件了。实际上， ASMFD 也使用了 UDEV 技术来绑定磁盘。下面验证合法的 I/O 操作，代码如下：

```
[root@18cASM rules.d]# pwd
```

```
/etc/udev/rules.d
[root@18cASM rules.d]# mv 99-oracle-asmdevices.rules
99-oracle-asmdevices.rules.bak
[root@18cASM rules.d]# ls
53-afd.rules  55-usm.rules  70-persistent-ipoib.rules
99-oracle-asmdevices.rules.bak
[root@18cASM rules.d]# cat 53-afd.rules
#
# AFD devices
KERNEL=="oracleafd/.*", OWNER="grid", GROUP="asmadmin",
MODE="0775"
KERNEL=="oracleafd/*", OWNER="grid", GROUP="asmadmin",
MODE="0775"
KERNEL=="oracleafd/disks/*", OWNER="grid", GROUP="asmadmin",
MODE="0664"
[root@18cASM rules.d]#
```

可以看到，磁盘是使用 UDEV 方式绑定磁盘路径的，代码如下：

```
[root@18cASM rules.d]# ll /dev/oracleafd/disks/
total 52
-rw-r--r-- 1 grid asmadmin 9 Sep 11 00:12 DATA1
-rw-r--r-- 1 grid asmadmin 9 Sep 11 00:12 DATA2
-rw-r--r-- 1 grid asmadmin 9 Sep 11 00:12 MIGR
-rw-r--r-- 1 grid asmadmin 9 Sep 11 00:12 OCR1
-rw-r--r-- 1 grid asmadmin 9 Sep 11 00:12 OCR2
-rw-r--r-- 1 grid asmadmin 9 Sep 11 00:12 OCR3
-rw-r--r-- 1 grid asmadmin 9 Sep 11 00:12 TESTDT1
-rw-r--r-- 1 grid asmadmin 9 Sep 11 00:12 TESTDT2
-rw-r--r-- 1 grid asmadmin 9 Sep 11 00:12 TESTDT3
-rw-r--r-- 1 grid asmadmin 9 Sep 11 00:12 TESTDT4
-rw-r--r-- 1 grid asmadmin 9 Sep 11 00:12 TESTDT5
-rw-r--r-- 1 grid asmadmin 9 Sep 11 00:12 TESTDT6
-rw-r--r-- 1 grid asmadmin 9 Sep 11 00:12 TESTDT7
```

（9）修改 Disk String。因为 ASM Disk String 参数中还有之前磁盘的信息，所以查询 v$asm_disk 视图会显示双重的信息，代码如下：

```
SQL> select name,label,path from v$asm_disk order by 3;
NAME                LABEL                    PATH
------------------------------- -------------------------------
                    OCR1                     /dev/asmdiskb
                    OCR2                     /dev/asmdiskc
                    OCR3                     /dev/asmdiskd
```

```
                     MIGR                   /dev/asmdiske
                     DATA2                  /dev/asmdiskf
                     DATA1                  /dev/asmdiskg
                     TESTDT7                     /dev/asmdiskh
                     TESTDT6                     /dev/asmdiski
                     TESTDT5                     /dev/asmdiskj
                     TESTDT4                     /dev/asmdiskk
                     TESTDT3                     /dev/asmdiskl
                     TESTDT1                     /dev/asmdiskm
                     TESTDT2                     /dev/asmdiskn
                     DATA1                  AFD:DATA1
                     DATA2                  AFD:DATA2
                     MIGR                   AFD:MIGR
                     OCR1                   AFD:OCR1
OCRVD_0001                OCR2                   AFD:OCR2
OCRVD_0002                OCR3                   AFD:OCR3
                     TESTDT1                     AFD:TESTDT1
                     TESTDT2                     AFD:TESTDT2
                     TESTDT3                     AFD:TESTDT3
                     TESTDT4                     AFD:TESTDT4
```

修改 Disk String 参数，使其只显示 ASMFD，代码如下：

```
[grid@18cASM ~]$ asmcmd dsset 'AFD:*'
```

再次查看 V$ASM_DISK 视图，就不会显示其他信息了，代码如下：

```
SQL> select name,label,path from v$asm_disk order by 3;

NAME                 LABEL                  PATH
------------------------------- ------------------------------
                     DATA1                  AFD:DATA1
                     DATA2                  AFD:DATA2
                     MIGR                   AFD:MIGR
                     OCR1                   AFD:OCR1
OCRVD_0001                OCR2                   AFD:OCR2
OCRVD_0002                OCR3                   AFD:OCR3
                     TESTDT1                     AFD:TESTDT1
                     TESTDT2                     AFD:TESTDT2
                     TESTDT3                     AFD:TESTDT3
                     TESTDT4                     AFD:TESTDT4
                     TESTDT5                     AFD:TESTDT5
                     TESTDT6                     AFD:TESTDT6
                        TESTDT7                  AFD:TESTDT7
```

（10）重启操作系统，所有磁盘将由 ASMFD 全权接管，被绑定的磁盘也会变成底层的磁盘路径，代码如下：

```
[grid@18cASM ~]$ ls -l /dev/disk/by-label/
total 0
lrwxrwxrwx 1 root root 9 Sep 11 00:12 DATA1 -> ../../sdg
lrwxrwxrwx 1 root root 9 Sep 11 00:12 DATA2 -> ../../sdf
lrwxrwxrwx 1 root root 9 Sep 11 00:12 MIGR -> ../../sde
lrwxrwxrwx 1 root root 9 Sep 11 00:12 OCR1 -> ../../sdb
lrwxrwxrwx 1 root root 9 Sep 11 00:12 OCR2 -> ../../sdc
lrwxrwxrwx 1 root root 9 Sep 11 00:12 OCR3 -> ../../sdd
lrwxrwxrwx 1 root root 9 Sep 11 00:12 TESTDT1 -> ../../sdm
lrwxrwxrwx 1 root root 9 Sep 11 00:12 TESTDT2 -> ../../sdn
lrwxrwxrwx 1 root root 9 Sep 11 00:12 TESTDT3 -> ../../sdl
lrwxrwxrwx 1 root root 9 Sep 11 00:12 TESTDT4 -> ../../sdk
lrwxrwxrwx 1 root root 9 Sep 11 00:12 TESTDT5 -> ../../sdj
lrwxrwxrwx 1 root root 9 Sep 11 00:12 TESTDT6 -> ../../sdi
lrwxrwxrwx 1 root root 9 Sep 11 00:12 TESTDT7 -> ../../sdh
```

10.3 ASM 的 I/O Filter 功能

ASM 的 I/O Filter 功能可以拒绝所有无效的 I/O 请求，防止意外覆写 ASM 磁盘的底层盘，比如直接在操作系统级别对磁盘使用 dd 命令操作，这种过滤对 root 用户的操作都生效。

（1）查看是否启用了 Filter，代码如下：

```
    [grid@18cASM ~]$ asmcmd help afd_filter
    afd_filter
          Sets the AFD filtering mode on a given disk path.
          If the command is executed without specifying a disk path
then filtering is set at node level.
    Synopsis
          afd_filter {-e | -d } [<disk-path>] [--all]
    Description
          The options for afd_filter are described below
          -e     - enable  AFD filtering mode
          -d     - disable AFD filtering mode
          --all  - set clusterwide AFD filtering mode
```

通过命令帮助可以看到，-e 是启用，-d 是禁用。查询结果 ENABLED 表示启用了 Filter，代码如下：

```
[grid@18cASM ~]$ asmcmd afd_lsdsk
--------------------------------------------------------------
Label                  Filtering  Path
==============================================================
DATA1                   ENABLED  /dev/sdg
DATA2                   ENABLED  /dev/sdf
MIGR                    ENABLED  /dev/sde
OCR1                    ENABLED  /dev/sdb
OCR2                    ENABLED  /dev/sdc
```

（2）创建测试磁盘组。

为了避免破坏测试环境，要创建一个测试磁盘组，代码如下：

```
[grid@18cASM ~]$ sqlplus / as sysasm
SQL*Plus: Release 18.0.0.0.0 - Production on Tue Sep 11 00:45:10
2018
Version 18.3.0.0.0
Copyright (c) 1982, 2018, Oracle. All rights reserved.
Connected to:
Oracle Database 18c Enterprise Edition Release 18.0.0.0.0 -
Production
Version 18.3.0.0.0

SQL> create diskgroup afdtest external redundancy disk
'AFD:TESTDT1,AFD:TESTDT2';
Diskgroup created.

[grid@18cASM ~]$ asmcmd lsdg
State    Type    Rebal  Sector  Logical_Sector  Block     AU
Total_MB  Free_MB  Req_mir_free_MB  Usable_file_MB  Offline_disks
Voting_files  Name
MOUNTED  EXTERN  N       512            512   4096  1048576
2048    1994             0          1994            0          N
AFDTEST/
MOUNTED  NORMAL  N       512            512   4096  4194304
2048    1816          1024           396            1          N
OCRVD/
```

（3）用 dd 命令写磁盘，代码如下：

```
[root@18cASM ~]# dd if=/dev/zero of=/dev/sdm
dd: writing to '/dev/sdm': No space left on device
2097153+0 records in
2097152+0 records out
1073741824 bytes (1.1 GB) copied, 41.0384 s, 26.2 MB/s
```

（4）将磁盘组卸载，再挂载，代码如下：

```
    [grid@18cASM ~]$ asmcmd umount afdtest
    [grid@18cASM ~]$ asmcmd mount afdtest
    [grid@18cASM ~]$ asmcmd lsdg
    State    Type    Rebal  Sector  Logical_Sector  Block       AU
Total_MB  Free_MB  Req_mir_free_MB  Usable_file_MB  Offline_disks
Voting_files  Name
    MOUNTED  EXTERN  N        512            512   4096  1048576
2048    1994            0          1994           0          N
AFDTEST/
    MOUNTED  NORMAL  N        512            512   4096  4194304
2048    1816          1024           396           1          N
OCRVD/
    [grid@18cASM ~]$
```

可以看到，虽然对整个磁盘使用 dd 命令进行了操作，但似乎并没有对 ASM 产生影响。

查看操作系统的日志文件/var/log/message，还可以发现其他一些信息，代码如下：

```
    Sep 11 00:48:23 18cASM kernel: Buffer I/O error on dev sdm, logical
block 13312, lost async page write
    Sep 11 00:48:23 18cASM kernel: Buffer I/O error on dev sdm, logical
block 13313, lost async page write
    Sep 11 00:48:23 18cASM kernel: Buffer I/O error on dev sdm, logical
block 13314, lost async page write
    Sep 11 00:48:23 18cASM kernel: Buffer I/O error on dev sdm, logical
block 13315, lost async page write
    Sep 11 00:48:23 18cASM kernel: Buffer I/O error on dev sdm, logical
block 13316, lost async page write
    Sep 11 00:48:23 18cASM kernel: Buffer I/O error on dev sdm, logical
block 13317, lost async page write
    Sep 11 00:48:23 18cASM kernel: Buffer I/O error on dev sdm, logical
block 13318, lost async page write
    Sep 11 00:48:23 18cASM kernel: Buffer I/O error on dev sdm, logical
block 13319, lost async page write
    Sep 11 00:48:23 18cASM kernel: Buffer I/O error on dev sdm, logical
block 13320, lost async page write
    Sep 11 00:48:23 18cASM kernel: Buffer I/O error on dev sdm, logical
block 13321, lost async page write
    Sep 11 00:48:54 18cASM kernel: buffer_io_error: 42618 callbacks
suppressed
```

（5）禁用 I/O Filter 后再测试。

禁用 I/O Filter 的代码如下：

```
[root@18cASM ~]# asmcmd afd_filter -d /dev/sdm
```

这里只对/dev/sdm 进行了禁用，注意这里用的是 root 账户，使用其他账户会报权限不足，代码如下：

```
    [grid@18cASM ~]$ asmcmd afd_lsdsk
--------------------------------------------------------------
Label                    Filtering  Path
==============================================================
TESTDT1                   DISABLED   /dev/sdm
TESTDT2                    ENABLED  /dev/sdn
TESTDT3                    ENABLED  /dev/sdl
TESTDT4                    ENABLED  /dev/sdk
```

同样，对/dev/sdm 使用 dd 命令后，磁盘就不能进行正常的挂载了，代码如下：

```
[root@18cASM ~]# dd if=/dev/zero of=/dev/sdm
dd: writing to '/dev/sdm': No space left on device
2097153+0 records in
2097152+0 records out
1073741824 bytes (1.1 GB) copied, 48.1095 s, 22.3 MB/s
    [grid@18cASM ~]$ asmcmd umount afdtest
[grid@18cASM ~]$ asmcmd mount afdtest
ORA-15032: not all alterations performed
ORA-15017: diskgroup "AFDTEST" cannot be mounted
ORA-15040: diskgroup is incomplete (DBD ERROR: OCIStmtExecute)
```

10.4　卸载 ASMFD

如果不想再使用 ASMFD，则可以通过如下步骤来卸载。注意，卸载后需要重启 CRS。

（1）修改 Disk String。

如果 Disk String 中没有之前配置的磁盘的路径信息，那么需要将之前配置的磁盘的路径信息加到其中（ASM_DISKSTRING），代码如下：

```
$ asmcmd dsset '/dev/asm*,AFD:*'
```

（2）查看当前集群中的所有节点及角色，代码如下：

```
$ olsnodes -a
```

（3）在集群中的所有节点执行下面的操作，注意操作用户（$和#的区别）。

关闭 CRS，代码如下：

```
# $ORACLE_HOME/bin/crsctl stop crs
```

关闭 ACFS，代码如下：

```
# $ORACLE_HOME/bin/acfsload stop
```

卸载 ASMFD，代码如下：

```
# $ORACLE_HOME/bin/asmcmd afd_deconfigure
```

启动 ACFS，代码如下：

```
# $ORACLE_HOME/bin/acfsload start
```

启动 CRS，代码如下：

```
# $ORACLE_HOME/bin/crsctl start crs
```

检查 ASMFD 的状态，代码如下：

```
$ $ORACLE_HOME/bin/asmcmd afd_state
```

（4）更新 Disk String，将 AFD 磁盘信息从 Disk String 中去掉，代码如下：

```
$ asmcmd dsset '/dev/asm*'
$ $ORACLE_HOME/bin/asmcmd dsget
```

第11章 Oracle Flex ASM

Oracle 从 12c 开始引入 Flex ASM 功能，Flex ASM 能够使数据库与集群中其他节点上的 ASM 实例一起运行。例如，有一个两节点的 RAC，其中一个节点的 ASM 实例挂掉了，那么该节点上的数据库可以切换到另一个节点的 ASM 实例上进行通信。在 Oracle 18c 之前，Flex ASM 是一个可选项，但在 Oracle 18c 中，安装界面上已经不提供复选框了，默认使用 Flex ASM。

查看是否启用了 Flex ASM，代码如下：

```
$ asmcmd showclustermode
ASM cluster : Flex mode enabled - Direct Storage Access
[grid@rac2 ~]$ crsctl get cluster mode status
Cluster is running in "flex" mode
```

查看 Flex ASM 集群的连接信息，代码如下：

```
SQL> SELECT instance_name, db_name, status FROM V$ASM_CLIENT;
INSTANCE_NAME   DB_NAME  STATUS
------------------------ ----------------- ------------------
+ASM1          +ASM    CONNECTED
orcl1          orcl    CONNECTED
orcl2          orcl    CONNECTED

$ asmcmd lsct data
DB_Name  Status    Software_Version  Compatible_version
Instance_Name  Disk_Group
+ASM    CONNECTED     18.0.0.0.0     18.0.0.0.0  +ASM       DATA
orcl    CONNECTED     18.0.0.0.0     18.0.0.0.0  orcl1      DATA
orcl    CONNECTED     18.0.0.0.0     18.0.0.0.0  orcl2      DATA
```

11.1 Flex ASM 高可用测试

Flex ASM 最重要的特点就是数据库不再依赖本节点的 ASM 是否正常运行。如果某个节点的 ASM 异常关闭，那么该节点上的数据库会自动连接到其他节点的 ASM 上继续运行。

查看 Flex ASM 中的实例信息，代码如下：

```
[grid@rac2 ~]$ srvctl status asm -detail
ASM is running on rac1
ASM is enabled.
ASM instance +ASM1 is running on node rac1
Number of connected clients: 4
Client names: +APX1:+APX:RAC -MGMTDB:_mgmtdb:RAC
cndba1:cndba:RAC rac1:_OCR:RAC

[grid@rac2 ~]$ srvctl config asm -detail
ASM home: <CRS home>
Password file: +OCR_VOTING/orapwASM
ASM listener: LISTENER
ASM is enabled.
ASM is individually enabled on nodes:
ASM is individually disabled on nodes:
ASM instance count: 3
Cluster ASM listener: ASMNET1LSNR_ASM
```

下面模拟一个节点的 ASM 实例故障，查看该节点上的数据库状态。

（1）关闭节点 1 的 ASM 实例，查看节点 1 的数据库状态及进程，代码如下：

```
SQL>select instance_name,instance_number from gv$instance;

INSTANCE_NAME          INSTANCE_NUMBER
-------------------------------- ---------------
cndba1                     1
cndba2                     2

[grid@ rac1 ~]$ srvctl status database -d cndba -detail

Instance cndba1 is running on node rac1
Instance cndba1 is connected to ASM instance +ASM1
Instance cndba2 is running on node rac2
```

```
Instance cndba2 is connected to ASM instance +ASM2
[root@rac1 software]# ps -ef|grep pmon
oracle    6347     1  0 15:12 ?        00:00:00 ora_pmon_cndba1
root      9451  1808  0 15:21 pts/1    00:00:00 grep --color=auto
pmon
grid     16270     1  0 13:35 ?        00:00:00 mdb_pmon_-MGMTDB
grid     26646     1  0 12:43 ?        00:00:00 asm_pmon_+ASM1
```

停止节点 1 的 ASM 实例，代码如下：

```
[grid@rac1 ~]$ srvctl stop asm -node rac1 -stopoption abort -force
```

查看进程，可以看到 asm_pmon_+ASM1 进程没有了，代码如下：

```
[root@rac1 software]# ps -ef|grep pmon
oracle    6347     1  0 15:12 ?       00:00:00 ora_pmon_cndba1
root      9954  1808  0 15:23 pts/1  00:00:00 grep --color=auto pmon
grid     16270     1  0 13:35 ?       00:00:00 mdb_pmon_-MGMTDB
```

查看节点 1 的 ASM 实例状态，代码如下：

```
[grid@rac1 ~]$ srvctl status asm
ASM is running on rac2  --可以看到ASM只在RAC2上运行

[grid@rac1 ~]$ srvctl status database -d cndba -detail
Instance cndba1 is running on node rac1
Instance cndba1 is connected to ASM instance +ASM2  --连接到节点
2 的ASM实例上
Instance cndba2 is running on node rac2
Instance cndba2 is connected to ASM instance +ASM2

[grid@rac1 ~]$ srvctl status asm -detail
ASM is running on rac2
ASM is enabled.
ASM instance +ASM2 is running on node rac2
Number of connected clients: 6
Client names: +APX1:+APX:rac +APX2:+APX:rac -MGMTDB:_mgmtdb:rac
cndba1:cndba:rac cndba2:cndba:rac rac2:_OCR:rac
```

检查集群服务状态，代码如下：

```
[grid@rac1 ~]$ crsctl check cluster
CRS-4537: Cluster Ready Services is online
CRS-4529: Cluster Synchronization Services is online
CRS-4533: Event Manager is online
```

查看 ASM 和 RAC 数据的进程，代码如下：

```
[root@rac1 software]# ps -ef|grep pmon
oracle   6347    1  0 15:12 ?       00:00:00 ora_pmon_cndba1
root     9954  1808  0 15:23 pts/1   00:00:00 grep --color=auto
pmon
grid    16270    1  0 13:35 ?       00:00:00 mdb_pmon_-MGMTDB
```

数据库 pmon 进程还在，ASM 的 pmon 进程已经没有了。

（2）再次查看节点 1 的数据库状态，代码如下：

```
SQL> select instance_name,instance_number from gv$instance;
INSTANCE_NAME    INSTANCE_NUMBER
---------------- ---------------
cndba2          2
cndba1          1
```

可以看到，集群中任意一个节点上的 ASM 实例出现故障，都不会影响该节点上的数据库正常运行。

（3）启动节点 1 上的 ASM 实例，又会恢复原状，代码如下：

```
[grid@rac1 ~]$srvctl start asm -node rac1

[grid@rac1 ~]$ srvctl status asm -detail
ASM is running on rac1,rac2
ASM is enabled.
ASM instance +ASM1 is running on node rac1
Number of connected clients: 3
Client names: +APX1:+APX:rac cndba1:cndba:rac rac1:_OCR:rac
ASM instance +ASM2 is running on node rac2
Number of connected clients: 4
Client names: +APX2:+APX:rac -MGMTDB:_mgmtdb:rac
cndba2:cndba:rac rac2:_OCR:rac
[grid@rac1 ~]$

[grid@rac1 ~]$ srvctl status database -d cndba -detail
Instance cndba1 is running on node rac1
Instance cndba1 is connected to ASM instance +ASM1
Instance cndba2 is running on node rac2
Instance cndba2 is connected to ASM instance +ASM2
```

通过上面的示例可以验证任意一个节点的 ASM 实例被意外关闭/启动，都不会影响该节点上的数据库状态，极大地增强了 RAC 的高可用性。

11.2　Oracle Flex 集群

Oracle Flex 集群中的每个节点都属于单个 GI（Grid Infrastructure）集群，该体系结构集中了基于应用程序需求的资源部署策略，以考虑各种服务级别、负载、故障响应和恢复。

Oracle Flex 集群中包含两种类型的节点：Hub 节点和 Leaf 节点。Oracle Flex 集群中的 Hub 节点最多可以有 64 个。Leaf 节点的数量则更多，每个 Hub 节点支持 64 个 Leaf 节点（也就是最多 4096 个 Leaf 节点）。Hub 节点和 Leaf 节点可以为不同类型的应用程序提供服务。

- Hub 节点类似于普通 GI 节点，通过私有网络进行连接，可直接访问共享存储。Hub 节点上的数据库实例可以提供读写服务。
- Leaf 节点与普通 GI 节点不同，Leaf 节点不需要直接访问共享存储，而是通过 Hub 节点请求数据。Leaf 节点上的数据库实例只能提供只读服务。因此，Flex 集群中可以没有 Leaf 节点，但是必须有 Hub 节点，而且 Leaf 节点不能单独存在（必须依附于 Hub 节点）。

在 Oracle 18c 的 Flex 集群架构中，还有一个 Reader 节点的概念，它是运行在 Leaf 节点上的 RAC 数据库实例，只能提供只读访问。使用 Reader 节点的优点是，如果需要重新配置 Hub 节点，则这些实例不会受到影响。在 Reader 节点上运行的查询业务不受重新配置 Hub 节点的影响，并且可以继续为连接到 Reader 节点的客户端提供服务，只要它所连接的 Hub 节点没有被“踢”出集群。每个 Hub 节点最多可以支持 64 个 Reader 节点。

只读节点上的数据库与普通的数据库或 Hub 节点上运行的数据库主要有以下几点区别。

- 数据库只能以只读模式运行。
- 只读节点上的数据库不受 Hub 节点的影响，只要 Hub 节点没有被“踢”出集群。
- 由于只能以只读模式运行，所以无法执行 DML 或 DDL 操作。
- 可以通过以下命令查看节点的角色：

```
[grid@rac1 ~]$ crsctl get node role config
```

```
Node 'rac1' configured role is 'hub'
[grid@rac2 ~]$ crsctl get node role config
Node 'rac2' configured role is 'hub'
```

从上面的查询结果看，在 Oracle 18c 的环境中，默认都是 Hub 节点。但是需要注意，Oracle 从 11g 到 12c、再到 18c 的过程中一直在改动，虽然 Oracle 在 12c 中引入了 Hub/Leaf 结构，但到了 Oracle 18c，从官方的信息看，Leaf 架构已经被废弃，即 Leaf 节点可能在以后的版本中不再被支持，但是 Oracle18c 版本中还支持。另外，从实际情况来看，Leaf 节点存在的意义不是很大，不仅占用单独的服务器，而且只提供只读服务，浪费资源。所以，Flex 集群的最终发展方向可能还需要时间来检验，如果读者想测试这种 Hub/Leaf 架构，可以直接参考官方手册。

11.3 ASM Flex 磁盘组和 Extent 磁盘组

Oracle ASM Flex 磁盘组是可以支持 Oracle ASM 文件组的。Oracle ASM 文件组是数据库中的一组文件，支持在文件组或数据库级别进行管理。Oracle ASM Extent 磁盘组具有 ASM Flex 磁盘组的所有功能，尤其在扩展/伸展集群环境中具有更高的可用性。

11.3.1 ASM 文件组

Oracle ASM 文件组是一组共享同一组属性和特征的文件。ASM 文件组的主要优点是能够为共享同一个磁盘组的每个数据库提供不同的配置。ASM 文件组的属性集包括冗余、再平衡优先级、再平衡速度、客户端兼容性、条带化、配额组和访问控制。

下面是文件组的一些说明。

- 磁盘组可以包含多个文件组，但至少有一个文件组，即默认文件组。
- 必须是 Flex 或 Extent 类型的磁盘组才能包含文件组。
- 磁盘组可以存储属于多个数据库的文件，每个数据库都有一个单独的文件组。
- 数据库在一个磁盘组中只能有一个文件组。
- 数据库可以跨多个磁盘组，其中多个文件组要位于不同的磁盘组中。

- 一个文件组只能同时属于一个磁盘组。
- 一个文件组只能用于一个数据库、PDB、CDB、卷或集群。
- 一个文件组只能属于一个配额组。
- 自动创建的文件组会与通用配额组关联。
- 在创建数据库的 PDB 或 CDB 时，如果已有文件组的名称与数据库、PDB、CDB 的名称相同，那么这个文件组用于描述该数据库；否则为数据库、PDB、CDB 创建一个新的文件组。
- 由于在创建数据库时会自动创建文件组，所以在删除数据库时，会自动删除之前创建的文件组。但是，如果是手动创建的文件组，那么在删除数据库时不会自动删除该文件组，需要手动删除。

在图 11-1 中，磁盘组 DATA1 和磁盘组 DATA2 中的 CNDBAPDB 文件组只用于 CNDBAPDB，磁盘组 DATA1 和磁盘组 DATA2 中的 LEIPDB 文件组只用于 LEIPDB，磁盘组 DATA1 和磁盘组 DATA2 中的 SUYIPDB 文件组只用于 SUYIPDB。

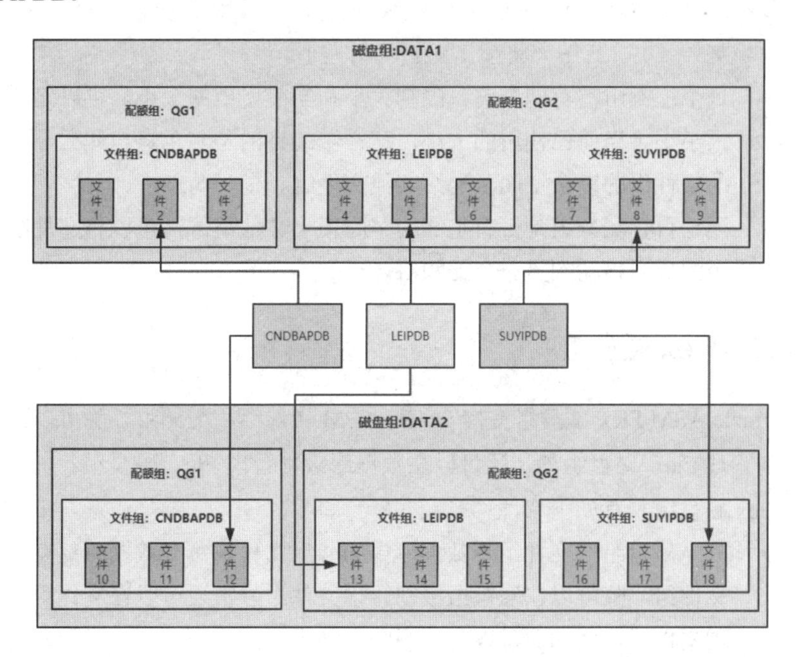

图 11-1　磁盘组和文件组示意图

磁盘组 DATA1 中的 CNDBAPDB 文件组属于磁盘组 DATA1 中的配额组

QG1。磁盘组 DATA2 中的 CNDBAPDB 文件组属于磁盘组 DATA2 中的配额组 QG1。磁盘组 DATA1 中的 LEIPDB 文件组和 SUYIPDB 文件组属于磁盘组 DATA1 中的配额组 QG2。磁盘组 DATA2 中的 LEIPDB 文件组和 SUYIPDB 文件组属于磁盘组 DATA2 中的配额组 QG2。

11.3.2 Oracle ASM 配额组

配额组是分配给一个文件组的配额。

下面是配额组的重要说明。

- 一个文件组只属于一个配额组。
- 一个配额组不能跨多个磁盘组。
- 配额组描述一个文件组或同一个磁盘组中的多个文件组使用的空间总和。
- 当在磁盘组中创建新文件并调整其大小时，会强制分配配额。
- 配额是物理空间，如果将配额限制设置为 10 MB，则双份镜像的 6 MB 文件（共 12MB）将超出配额限制大小。
- 每个配额组都有两个值：限制大小和当前使用的大小。可以设置限制大小小于当前已使用的大小，这样做的目的是防止将与此配额组相关的文件组的剩余空间分配给其他文件组。
- 无论目标配额组是否有足够的空间用于文件组，都可以将文件组从一个配额组移动到另一个配额组。

11.3.3 Flex 磁盘组

Oracle ASM Flex 磁盘组支持 Oracle ASM 文件组和配额组，除可以在磁盘组级别管理 Flex 磁盘组外，还可以在数据库级别管理 Flex 磁盘组。

Flex 磁盘组具有如下特性。

- Flex 磁盘组的文件组是数据库中的一组文件。每个数据库都有自己的文件组，除可以在磁盘组级别管理文件组外，还可以在文件组级别管理文件组。例如，可以为不同的文件组指定不同的冗余类型和再平衡设置，文件组与配额组关联。
- Flex 磁盘组的每个文件组都有自己的冗余级别。

- Flex 磁盘组和 High 冗余类型的磁盘组一样，至少需要 3 个故障组，并且同时允许两个故障组不可用。
- 可以将 Normal 和 High 冗余类型的磁盘组转换为 Flex 磁盘组，但在转换为 Flex 磁盘组之前，至少要有 3 个故障组，注意 External 冗余类型不支持转换。
- Flex 磁盘组可以创建基于时间点的数据库复制。
- 需要启用 Virtually Allocated Metadata（VAM）。
- 参数 COMPATIBLE.ASM 和 COMPATIBLE.RDBMS 的值必须设置为 12.2 或以上。
- V$ASM_DISKGROUP 视图中的 REQUIRED_MIRROR_FREE_MB 和 USABLE_FILE_MB 的列值不会显示 Flex 磁盘组的相关数据。

11.3.4 Extent 磁盘组

Extent 磁盘组具有所有 Flex 磁盘组的特性，多用于可扩展集群的高可用，该集群可包含跨多个不同物理站点的节点。

Extent 磁盘组具有如下特性。

- Extent 磁盘组中的每个文件组可以有自己的冗余配置。
- Extent 磁盘组中的文件和文件组的冗余级别是针对数据站点的，而不是针对磁盘组的，即数据副本数量等于参数 REDUNDANCY 的值乘以数据站点的数量。例如，将参数 REDUNDANCY 设置为 MIRROR，即指定两个副本，那么两个站点共有 4 个副本。
- Extent 磁盘组允许整个数据站点的数据丢失，以及另一个数据站点中最多两个故障组的数据丢失。
- 在创建磁盘组期间，所有数据站点必须具有相同数量的故障组。
- 为每个磁盘组指定配额组，而不是为每个数据站点指定配额组。
- 配额组的限制大小是所有数据站点中所有副本所需的物理空间。例如，有两个数据站点，冗余设置为 MIRROR 的 6 MB 文件使用 24 MB 的配额限制 （4 个副本）。
- Extent 磁盘组至少要有 3 个站点，最多支持 5 个站点：两个数据站点和一个仲裁站点才能创建 Extent 磁盘组。每个数据站点应有 3 个故障组，而仲裁站点只有一个故障组。

- 对于数据库文件，其冗余级别由 Extent 磁盘组的文件组的冗余级别决定。
- 参数 COMPATIBLE.ASM 和 COMPATIBLE.RDBMS 的值必须设置为 12.2 或更高。
- 不支持从其他磁盘组类型转换为 Extent 磁盘组。

11.3.5 相关操作示例

1. 查看磁盘的相关信息

查看磁盘的相关信息，代码如下：

```
SQL> set lines 130
SQL> set pages 200
SQL> col path for a30
SQL> select name,label,path from v$asm_disk order by 3;

NAME                 LABEL                    PATH
------------------------------ ------------------------------
------------------------------
DATA1                DATA1                    AFD:DATA1
DATA2                DATA2                    AFD:DATA2
                     MIGR                     AFD:MIGR
OCR1                 OCR1                     AFD:OCR1
OCRVD_0001           OCR2                     AFD:OCR2
OCRVD_0002           OCR3                     AFD:OCR3
                     TESTDT1                  AFD:TESTDT1
                     TESTDT2                  AFD:TESTDT2
                     TESTDT3                  AFD:TESTDT3
                     TESTDT4                  AFD:TESTDT4
                     TESTDT5                  AFD:TESTDT5
                     TESTDT6                  AFD:TESTDT6
                     TESTDT7                  AFD:TESTDT7
```

在当前环境中，TESTDT1～TESTDT7 都没有被使用，可以创建 Flex 磁盘组，进行相关的测试。

2. 创建具有 3 个故障组的 Normal 冗余磁盘组

创建具有 3 个故障组的 Normal 冗余磁盘组，代码如下：

```
SQL> CREATE DISKGROUP CNDBA NORMAL REDUNDANCY
```

```
    failgroup fg1 disk 'AFD:TESTDT1'
    failgroup fg2 disk 'AFD:TESTDT2'
    failgroup fg3 disk 'AFD:TESTDT3'
    ATTRIBUTE 'au_size'='4M', 'compatible.asm' =
'18.0','compatible.rdbms' = '18.0','compatible.advm' = '18.0';
    Diskgroup created.
```

验证磁盘组的信息，代码如下：

```
    SQL> set lines 120
    SQL> col name for a15
    SQL> col name for a15
    SQL> col failgroup for a15
    SQL> col label for a15
    SQL> col path for a15
    SQL> select b.name,b.type,a.name,a.failgroup,a.label,a.path from
v$asm_disk a, v$asm_diskgroup b where a.group_number=b.group_number;

    NAME     TYPE   NAME         FAILGROUP      LABEL          PATH
    --------------- ------ --------------- ---------------
--------------- ---------------
    OCRVD    NORMAL OCR1         OCR1           OCR1           AFD:OCR1
    OCRVD    NORMAL OCRVD_0001      OCRVD_0001      OCR2
AFD:OCR2
    OCRVD    NORMAL OCRVD_0002      OCRVD_0002      OCR3
AFD:OCR3
    DATA     NORMAL DATA2        DATA2          DATA2          AFD:DATA2
    DATA     NORMAL DATA1        DATA1          DATA1          AFD:DATA1
    CNDBA    NORMAL TESTDT3         FG3            TESTDT3
AFD:TESTDT3
    CNDBA    NORMAL TESTDT1         FG1            TESTDT1
AFD:TESTDT1
    CNDBA    NORMAL TESTDT2         FG2            TESTDT2
AFD:TESTDT2

    8 rows selected.
```

将 Normal 类型的磁盘组 CNDBA 转换成 Flex 磁盘组，代码如下：

```
    SQL> alter diskgroup cndba convert redundancy to flex;
    Diskgroup altered.
    SQL>select b.name,b.type,a.name,a.failgroup,a.label,a.path from
v$asm_disk a, v$asm_diskgroup b where a.group_number=b.group_number;
```

```
NAME     TYPE   NAME         FAILGROUP     LABEL          PATH
------------- --------------- --------------- ---------------
OCRVD        NORMAL OCR1        OCR1         OCR1
AFD:OCR1
OCRVD        NORMAL OCRVD_0001     OCRVD_0001     OCR2
AFD:OCR2
OCRVD        NORMAL OCRVD_0002     OCRVD_0002     OCR3
AFD:OCR3
DATA     NORMAL DATA2       DATA2        DATA2        AFD:DATA2
DATA     NORMAL DATA1       DATA1        DATA1        AFD:DATA1
CNDBA        FLEX   TESTDT3         FG3          TESTDT3
AFD:TESTDT3
CNDBA        FLEX   TESTDT1         FG1          TESTDT1
AFD:TESTDT1
CNDBA        FLEX   TESTDT2         FG2          TESTDT2
AFD:TESTDT2

8 rows selected.
```

之前是从 Normal 类型转换成 Flex 磁盘组的，下面将 CNDBA 磁盘组删除（Drop），然后直接创建 Flex 磁盘组，代码如下：

```
SQL> drop diskgroup cndba including contents;
Diskgroup dropped.
SQL> CREATE DISKGROUP CNDBA FLEX REDUNDANCY
failgroup fg1 disk 'AFD:TESTDT1'
failgroup fg2 disk 'AFD:TESTDT2'
failgroup fg3 disk 'AFD:TESTDT3'
ATTRIBUTE 'au_size'='4M', 'compatible.asm' =
'18.0','compatible.rdbms' = '18.0','compatible.advm' = '18.0';
Diskgroup created.
SQL> select b.name,b.type,a.name,a.failgroup,a.label,a.path from
v$asm_disk a, v$asm_diskgroup b where a.group_number=b.group_number;

NAME     TYPE   NAME         FAILGROUP     LABEL          PATH
--------------- --------------- --------------- ---------------
OCRVD    NORMAL OCR1           OCR1         OCR1         AFD:OCR1
OCRVD    NORMAL OCRVD_0001     OCRVD_0001   OCR2         AFD:OCR2
OCRVD    NORMAL OCRVD_0002     OCRVD_0002   OCR3         AFD:OCR3
DATA     NORMAL DATA2          DATA2        DATA2        AFD:DATA2
DATA     NORMAL DATA1          DATA1        DATA1        AFD:DATA1
CNDBA    FLEX   TESTDT3         FG3         TESTDT3      AFD:TESTDT3
```

```
CNDBA    FLEX   TESTDT1          FG1      TESTDT1       AFD:TESTDT1
CNDBA    FLEX   TESTDT2          FG2      TESTDT2       AFD:TESTDT2

8 rows selected.
```

这里要注意一点，Flex 磁盘组的参数 required_mirror_free_mb 和 useable_file_mb 的值显示为 0 是正常现象，代码如下：

```
[grid@18cASM ~]$ asmcmd lsdg
State    Type    Rebal  Sector  Logical_Sector  Block       AU
Total_MB  Free_MB  Req_mir_free_MB  Usable_file_MB  Offline_disks
Voting_files  Name
MOUNTED  FLEX    N         512            512   4096  4194304
3072     2760              0           0            0           N  CNDBA/
MOUNTED  NORMAL  N         512            512   4096  4194304
20480    11800             0        5900            0           N  DATA/
MOUNTED  NORMAL  N         512            512   4096  4194304
3072     2716           1024         846            0           N  OCRVD/
```

3. 向 CNDBA 磁盘组中添加配额组 QG

向 CNDBA 磁盘组中添加配额组 QG，代码如下：

```
SQL> alter diskgroup cndba add quotagroup qg set 'quota'= 10m;
Diskgroup altered.
```

可以通过 V$ASM_QUOTAGROUP 视图查看配额组的相关信息（如图 11-2 所示），代码如下：

```
SELECT t1.NAME AS GROUP_NAME,
       t2.QUOTAGROUP_NUMBER,
       t2.NAME AS QUOTA_NAME,
       t2.USED_QUOTA_MB,
       t2.QUOTA_LIMIT_MB,
       F.NAME AS FILE_NAME,
       F.CLIENT_NAME,
       F.USED_QUOTA_MB
  FROM V$ASM_DISKGROUP t1
       LEFT JOIN V$ASM_QUOTAGROUP t2 ON t1.GROUP_NUMBER =
t2.GROUP_NUMBER
       LEFT JOIN V$ASM_FILEGROUP F
         ON    F.QUOTAGROUP_NUMBER = t2.QUOTAGROUP_NUMBER
           AND F.GROUP_NUMBER = t2.GROUP_NUMBER
ORDER BY GROUP_NAME, QUOTAGROUP_NUMBER;
```

GROUP_NAME	QUOTAGROUP_NUMBER	QUOTA_NAME	USED_QUOTA_MB	QUOTA_LIMIT_MB	FILE_NAME	CLIENT_NAME	USED_QUOTA_MB_1
CNDBA	1	GENERIC	0	0	DEFAULT_FILEGROUP		0
CNDBA	2	QG	0	10			
DATA							
OCRVD							

图 11-2　查看配额组的相关信息

从图 11-2 可以看出，CNDBA 磁盘组默认有一个名字为 GENERIC 的配额组和一个默认名字为 DEFAULT_FILEGROUP 的文件组。我们创建了 QG，大小是 10MB。配额组的相关属性都是可修改的，比如把 QG 的配额改成 20MB，命令如下：

```
SQL> alter diskgroup cndba modify quotagroup qg set 'quota'= 20m;
Diskgroup altered.
```

4. 向 CNDBA 磁盘组添加文件组

在创建 Flex 磁盘组时会生成一个默认的文件组。在前面讲解理论部分时也提过，一个文件组只能与一个实例相对应，可以使用默认的文件组，也可以在数据库和文件组关联时，创建新的文件组。

向 CNDBA 磁盘组添加文件组的代码如下：

```
SQL> show pdbs
   CON_ID CON_NAME                 OPEN MODE  RESTRICTED
---------- ------------------------------ ---------- ----------
     2 PDB$SEED             READ ONLY  NO
     3 DAVE             READ WRITE NO
SQL> alter diskgroup cndba add filegroup file_cndba database dave
set 'quota_group' = 'qg';
Diskgroup altered.
SQL> select group_number,filegroup_number,name,client_name from
v$asm_filegroup ;
GROUP_NUMBER FILEGROUP_NUMBER NAME         CLIENT_NAM
------------ ---------------- ------------------------------
----------
       3        0 DEFAULT_FILEGROUP
       3        1 FILE_CNDBA            DAVE
```

Flex ASM 磁盘组可以被设置为任意保护模式（3 副本、2 副本、无保护），默认情况下，Flex 冗余度的磁盘组使用 2 副本模式。磁盘组本身也有很多属性，可以通过 v$asm_file 视图查看，这里直接使用 asmcmd 命令查看，代码如下：

```
[grid@18cASM ~]$ asmcmd lsfg -G cndba --filegroup file_cndba
File Group  Disk Group  Property          Value   File Type
FILE_CNDBA  CNDBA      PRIORITY          MEDIUM
FILE_CNDBA  CNDBA      COMPATIBLE.CLIENT
FILE_CNDBA  CNDBA      REDUNDANCY         HIGH   CONTROLFILE
FILE_CNDBA  CNDBA      STRIPING          FINE   CONTROLFILE
FILE_CNDBA  CNDBA      REDUNDANCY         MIRROR  DATAFILE
FILE_CNDBA  CNDBA      STRIPING          COARSE  DATAFILE
FILE_CNDBA  CNDBA      REDUNDANCY         MIRROR  ONLINELOG
FILE_CNDBA  CNDBA      STRIPING          COARSE  ONLINELOG
FILE_CNDBA  CNDBA      REDUNDANCY         MIRROR  ARCHIVELOG
FILE_CNDBA  CNDBA      STRIPING          COARSE  ARCHIVELOG
FILE_CNDBA  CNDBA      REDUNDANCY         MIRROR  TEMPFILE
FILE_CNDBA  CNDBA      STRIPING          COARSE  TEMPFILE
FILE_CNDBA  CNDBA      REDUNDANCY         MIRROR  BACKUPSET
```

这些属性也可以直接进行修改，代码如下：

```
SQL> alter diskgroup cndba modify filegroup file_cndba set
'datafile.redundancy'='high';
Diskgroup altered.
```

5. 删除文件组

删除文件组的代码如下：

```
SQL> alter diskgroup cndba drop filegroup file_cndba cascade;
Diskgroup altered.
```

简单总结一下配额组、文件组和数据库实例的关系：配额组是控制空间容量使用的，文件组是控制文件属性的，只有在 Flex 或 Extent 磁盘组上才能创建这两种对象；文件组与数据库实例相对应，通过这种架构可以控制同一个磁盘组内不同数据库实例的文件属性和使用空间的大小。

6. 创建 Extent 磁盘组

前面提过，Extent 磁盘组由 3 个站点组成，其中有两个是数据站点，每个数据站点内有 3 个故障组，另外一个是仲裁站点，仲裁站点内只有一个故障组。对于目前的架构，容灾都在同一个机房，这种架构可以规避单点故障。但作为一个新技术，到最终大规模的使用，中间还需要一些时间。下面简单模拟一下创建过程，其他的不再进行测试。

根据官方手册，创建 Extent 磁盘组的代码如下：

```
SQL>CREATE DISKGROUP CNDBA EXTENDED REDUNDANCY
      SITE HEFEI  FAILGROUP fg1 DISK 'AFD:TESTDT1'
               FAILGROUP fg2 DISK 'AFD:TESTDT2'
               FAILGROUP fg3 DISK 'AFD:TESTDT3'
      SITE ANQING FAILGROUP fg4 DISK 'AFD:TESTDT4'
               FAILGROUP fg5 DISK 'AFD:TESTDT5'
               FAILGROUP fg6 DISK 'AFD:TESTDT6'
      SITE QM QUORUM
               FAILGROUP fg7 DISK 'AFD:TESTDT7';
```

在执行以上代码时会报如下错误：

```
ORA-59712: The site identified by site name 'ANQING' does not exist
```

尝试使用 crsctl 命令进行创建，代码如下：

```
[grid@18cASM ~]$ crsctl add cluster  -h
Usage:
  crsctl add cluster site <site_name> [-guid <site_guid>]
    Add the site to the cluster.
Where
     site_name     The site name of the new site
     site_guid     The site GUID (global unique ID) of the new
site

[grid@rac1 ~]$ crsctl add cluster site hefei
CRS-6592: Site 'hefei' cannot be added, deleted or quarantined
because the cluster is not operating in extended mode.
CRS-4000: Command Add failed, or completed with errors.
```

这时提示集群不是 Extended 模式。翻看一下安装手册，发现其中有一项配置（Gonfigure as an Oracle Extended cluster），如图 11-3 所示。

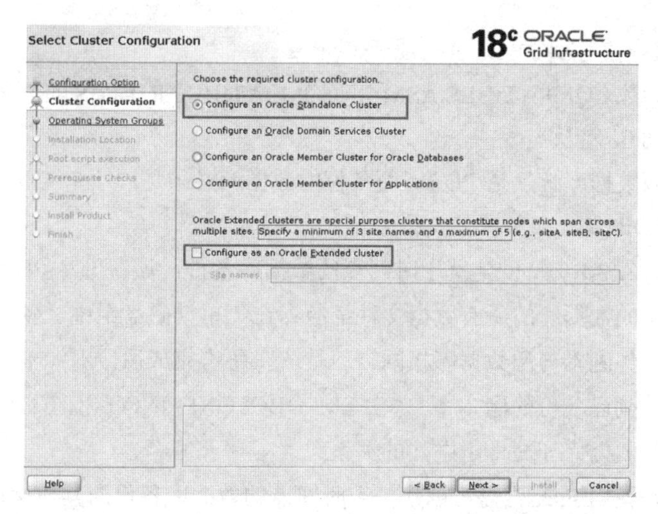

图 11-3　安装手册

在 ASMCA 工具中选择磁盘时，也会多出一列“Site”，如图 11-4 所示。

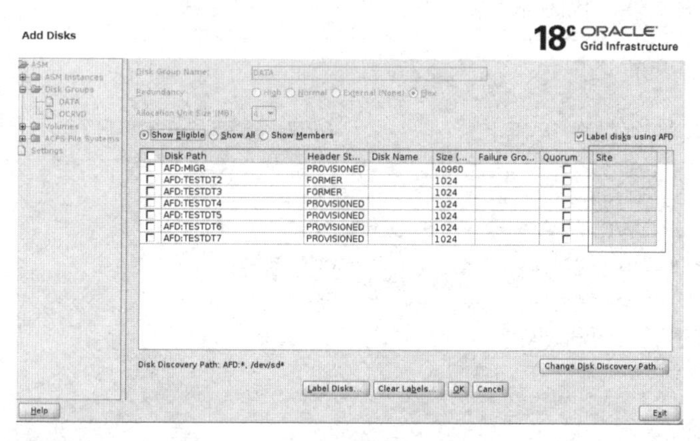

图 11-4　在 ASMCA 工具中选择磁盘时多出一列“Site”

所以，要使用 Extent 磁盘组，在安装 RAC 时必须选择 Extended 模式。

11.4　使用 Flex 磁盘组创建基于时间点的数据库备份

在使用 Flex 磁盘组创建基于时间点的数据库备份时，必须满足以下条件。

- 必须是 Extent 或 Flex 磁盘组。

- 数据库版本是 Oracle 18c 及以上。
- 参数 COMPATIBLE.ASM 和 COMPATIBLE.RDBMS 的值必须设置为 18.0 及以上。
- 源数据库必须是 PDB，而备份的数据库必须是同一个 CDB 中的其他 PDB。

Oracle ASM 支持创建基于时间点的数据库备份，该备份是 PDB 的镜像副本。要备份的数据库的所有数据文件都应存在于单个磁盘组中。备份过程中创建的数据文件副本与源数据库相同。但是，在备份完成后，对源数据库的更改操作不会传到已备份的副本中。V$ASM_DBCLONE_INFO 视图可以显示源数据库、备份数据库和文件组之间的关系。

下面是使用 Flex 磁盘组创建基于时间点的数据库备份的具体操作步骤。

（1）准备镜像副本

准备镜像副本的过程涉及创建备份文件并将其与源文件进行链接，但是不会复制数据文件，代码如下：

```
SQL> show pdbs

   CON_ID CON_NAME              OPEN MODE  RESTRICTED
---------- ------------------------------ ---------- ----------
     2 PDB$SEED           READ ONLY  NO
     3 DAVE            READ WRITE NO
SQL> alter session set container=dave;
Session altered.
SQL> show con_name
CON_NAME
------------------------------
DAVE
SQL> alter pluggable database prepare mirror copy dave_mirror;
Pluggable database altered.
```

查看视图是否准备好备份，代码如下：

```
SQL>select
dbclone_name,mirrorcopy_name,dbclone_status,parent_dbname,parent_f
ilegroup_name from v$asm_dbclone_info;
DBCLONE_NAME    MIRRORCOPY_NAME    DBCLONE_STATUS
PARENT_DBNAME       ENT_FILEGROUP_NAME
------------------------------------ -------------------------
```

```
    DB_UNKNOWN       DAVE_MIRROR    PREPARED           CNDBA_DAVE
CNDBA_DAVE
```

（2）分离镜像副本并创建数据库备份

步骤（1）完成后就可以通过备份的镜像 DAVE_MIRROR 来创建 PDB 了，代码如下：

```
SQL> show con_name
CON_NAME
------------------------------
CDB$ROOT
SQL> create pluggable database cpdave from dave using mirror copy
dave_mirror;
Pluggable database created.
SQL> show pdbs
   CON_ID CON_NAME               OPEN MODE  RESTRICTED
---------- ------------------------------ ---------- ----------
       2 PDB$SEED             READ ONLY  NO
       3 DAVE             READ WRITE NO
       4 CPDAVE               MOUNTED
SQL> alter pluggable database cpdave open;
Pluggable database altered.
SQL> show pdbs

   CON_ID CON_NAME               OPEN MODE  RESTRICTED
---------- ------------------------------ ---------- ----------
       2 PDB$SEED             READ ONLY  NO
       3 DAVE             READ WRITE NO
       4 CPDAVE               READ WRITE NO
```

还可以查询 V$ASM_DBCLONE_INFO 、 V$ASM_FILEGROUP 和 V$ASM_FILEGROUP_PROPERTY 视图来获取其他信息。注意下面语句中的状态信息，从开始的 PREPARED 到现在的 SPLIT COMPLETED，代码如下：

```
SQL>select
dbclone_name,mirrorcopy_name,dbclone_status,parent_dbname,parent_f
ilegroup_name from v$asm_dbclone_info;
    DBCLONE_NAME     MIRRORCOPY_NAME    DBCLONE_STATUS
PARENT_DBNAME        PARENT_FILEGROUP_NAME
    --------------- -------------------- -------------------------
    CNDBA_CPDAVE     DAVE_MIRROR    SPLIT COMPLETED     CNDBA_DAVE
CNDBA_DAVE
```

```
SQL> select name, group_number, filegroup_number from
v$asm_filegroup;

NAME         GROUP_NUMBER FILEGROUP_NUMBER
-------------------- ------------ ----------------
DEFAULT_FILEGROUP          1         0
CNDBA_DAVE                 1         1
DAVE_MIRROR                1         2
CNDBA_CDB$ROOT             1         3
```

（3）删除备份的镜像副本

镜像副本是可以被删除的，但删除的前提是还没有执行分离镜像的操作，如果执行了，就会报错，代码如下：

```
SQL> alter session set container=dave;
Session altered.
SQL> alter pluggable database drop mirror copy dave_mirror;
alter pluggable database drop mirror copy dave_mirror
*
ERROR at line 1:
ORA-59024: dropping mirror copy 'DAVE_MIRROR' in disk group 'DATA'
failed as the file split is complete
SQL> alter pluggable database prepare mirror copy dave_mirror2;
Pluggable database altered.
SQL> alter pluggable database drop mirror copy dave_mirror2;
Pluggable database altered.
SQL>select
dbclone_name,mirrorcopy_name,dbclone_status,parent_dbname,parent_f
ilegroup_name from v$asm_dbclone_info;
DBCLONE_NAME    MIRRORCOPY_NAME    DBCLONE_STATUS
PARENT_DBNAME       PARENT_FILEGROUP_NAME
------------------------------ -------------------------
CNDBA_CPDAVE  DAVE_MIRROR  SPLIT COMPLETED  CNDBA_DAVE
CNDBA_DAVE
DB_UNKNOWN     DAVE_MIRROR2  DROPPING         CNDBA_DAVE
CNDBA_DAVE
```

可以看到，被删除的镜像的状态变成 DROPPING。当删除操作完成后，对应的文件组和其中的文件也会被删除。

第 12 章 Oracle RAC

12.1 Oracle RAC 概述

在单实例环境中，数据库和实例的关系是一对一的，而在 RAC 环境中，数据库和实例的关系是一对多的，即一个数据库可以有多个实例，最多可以有 100 个实例。

Oracle RAC 数据库体系结构与单实例体系结构不同，其特点如下：

- 每个实例有自己的 Redo 线程；
- 每个实例有自己的 UNDO 表空间。

与单实例相比，RAC 多节点可提供更高的吞吐量和更好的可扩展性（可以随时添加节点到集群中）。Oracle RAC 是通过集群软件将多个数据库整合为一个整体对外提供服务的，但是它们不共享任何硬件资源，所以提供的性能更好。

如图 12-1 所示是一个简单的 RAC 架构图，每个数据库实例运行在不同的服务器上。

通常情况下，Oracle RAC 集群中的所有节点都位于同一个数据中心，但是在 Oracle Extented 集群中，Oracle RAC 可以跨站点（节点可以分布在不同的数据中心），并且和普通 RAC 集群一样作为一个整体对外提供服务。在 Extented 集群中，集群中的各个节点在地理上通常是分开的，可以是不同城市之间的不同机房。在 Extented 集群中，Oracle 为了保证数据的高可用，要求数据存储在两个站点，因此需要对数据进行镜像存储。这种架构对于特殊情况（停电、自然灾害）可以提供更好的数据保护，但这种架构也有缺点，比如由于距离的原因，会导致延迟较长。

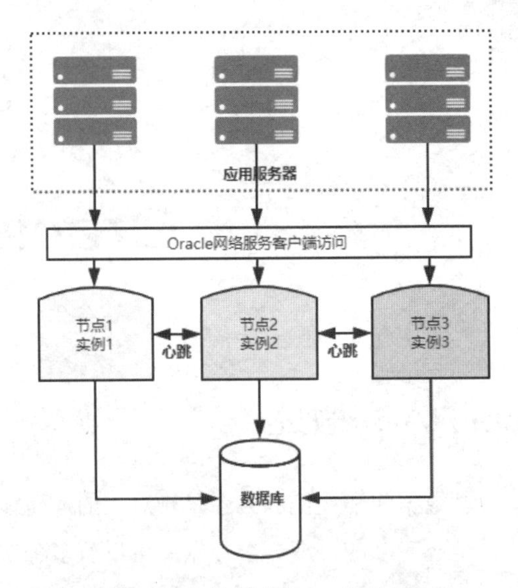

图 12-1　简单的 RAC 架构图

12.2　Oracle 集群软件

Oracle 集群软件是专为 Oracle RAC 设计的，集群软件注册并管理 RAC 数据库和其他必需组件，如 VIP、SCAN（Single Client Access Name，包括 SCAN VIP 和 SCAN 监听）、Oracle Notification Service 和 Oracle Net 监听等。这些资源在节点启动时自动启动，如果启动失败，则尝试自动重启。

Oracle 集群软件管理的所有对象统称为 CRS 资源。CRS 资源可以是数据库、实例、服务、监听、VIP 和进程。Oracle 集群软件根据存储在 OCR（Oracle Cluster Registry）中的资源配置信息管理 CRS 资源，也可以使用 srvctl 命令管理 CRS 资源。

12.3　Oracle RAC 后台进程

Oracle RAC 的后台进程很多，可以直接通过后台进程的 v$bgprocess 视图进行查看，代码如下：

```
SQL> select count (1) from v$bgprocess;
 COUNT (1)
----------
```

```
      354
SQL> select name, description from v$bgprocess where rownum<10;
NAME  DESCRIPTION
---------------------------------------------------------------
ABMR  Auto BMR Background Process
ACFS  ACFS CSS
ACMS  Atomic Controlfile to Memory Server
AQPC  AQ Process Coord
ARB0  ASM Rebalance 0
ARB1  ASM Rebalance 1
ARB2  ASM Rebalance 2
ARB3  ASM Rebalance 3
ARB4  ASM Rebalance 4
9 rows selected.
```

在 RAC 的众多后台进程中，最主要的同时也是保证 GI 正常工作的是 GCS（Global Cache Service）和 GES（Global Enqueue Service）两个进程，这两个进程和 GRD（Global Resource Directory）共同实现缓存融合（Cache Fusion）。下面对几个主要的后台进程进行说明。

- ACMS：实例的 ACMS 进程是一个代理，用于确保分布式 SGA 内存在更新成功时进行全局提交或在发生故障时全局中止。
- GTX0-j：为 Oracle RAC 中的 XA 全局事务提供透明支持。Oracle 根据 XA 全局事务的工作负载自动调整该进程的数量。
- LMON：监视集群中的全局队列和资源，并执行全局队列恢复操作。
- LMD：管理每个实例中来自远端的资源请求。
- LMS：通过在 GRD 中记录的信息来维护数据文件的状态和每个高速缓存块的记录。LMS 进程还控制传到远端实例的消息，并管理全局数据块访问，以及在不同实例缓冲区的高速缓存之间传输块镜像。该进程是缓存融合的一部分。
- LCK0：管理 Non-Cache Fusion 资源请求，例如库和行缓存请求。
- RMSn：为 Oracle RAC 执行可管理性任务，包括在将新实例添加到集群中时创建与 Oracle RAC 相关的资源。
- RSMN：管理远端实例上的后台 Slave 进程的创建和通信。这些后台 Slave 进程代表在另一个实例中运行的协调进程执行的任务。

这里要注意进程功能和进程缩写的对应关系，代码如下：

```
[grid@rac1 ~]$ ps -ef|grep acms
```

```
oracle    6873     1  0 Sep11 ?        00:00:01 ora_acms_cndba1
grid     18361 13765  0 09:56 pts/0    00:00:00 grep --color=auto acms
[grid@rac1 ~]$ ps -ef|grep gtx
oracle   10586     1  0 Sep11 ?        00:00:01 ora_gtx0_cndba1
grid     18650 13765  0 09:56 pts/0    00:00:00 grep --color=auto gtx
[grid@rac1 ~]$
[grid@rac1 ~]$ ps -ef|grep lmd
grid      5172     1  0 Sep11 ?        00:01:08 asm_lmd0_+ASM1
oracle    6882     1  0 Sep11 ?        00:01:06 ora_lmd0_cndba1
grid     18551 13765  0 09:56 pts/0    00:00:00 grep --color=auto lmd
[grid@rac1 ~]$ ps -ef|grep lms
grid      5174     1  0 Sep11 ?        00:02:59 asm_lms0_+ASM1
oracle    6884     1  0 Sep11 ?        00:02:20 ora_lms0_cndba1
grid     18587 13765  0 09:56 pts/0    00:00:00 grep --color=auto lms
```

从上面的代码看，部分进程的缩写和描述是对应的，比如 ACMS，但是 LMD 与 LMS 的缩写和功能描述是不对应的，需要读者特别留意一下。

12.4 Oracle 18c 中的新 CRS 资源

Oracle 软件版本在不断升级，很多方面都有细微的变化，下面简单介绍一下 Oracle 18c RAC 中新增的几个 CRS 资源。首先看一下 Oracle 11gR2 中的资源，代码如下：

```
[grid@rac1 ~]$ crsctl stat res -t
--------------------------------------------------------------------------------
NAME           TARGET  STATE        SERVER                   STATE_DETAILS
--------------------------------------------------------------------------------
Local Resources
--------------------------------------------------------------------------------
ora.ACFS.dg
               ONLINE  ONLINE       rac1
ora.DATA.dg
               ONLINE  ONLINE       rac1
ora.LISTENER.lsnr
               ONLINE  ONLINE       rac1
ora.OCR.dg
               ONLINE  ONLINE       rac1
ora.asm
               ONLINE  ONLINE       rac1                     Started
ora.gsd
               OFFLINE OFFLINE      rac1
```

```
ora.net1.network
            ONLINE  ONLINE      rac1
ora.ons
            ONLINE  ONLINE      rac1
ora.registry.acfs
            ONLINE  ONLINE      rac1
--------------------------------------------------------------------
Cluster Resources
--------------------------------------------------------------------
ora.LISTENER_SCAN1.lsnr
     1      ONLINE  ONLINE      rac1
ora.cndba.db
     1      OFFLINE OFFLINE                          Instance
Shutdown
     2      OFFLINE OFFLINE
ora.cvu
     1      ONLINE  ONLINE      rac1
ora.oc4j
     1      ONLINE  OFFLINE                          STARTING
ora.rac1.vip
     1      ONLINE  ONLINE      rac1
ora.rac2.vip
     1      ONLINE  INTERMEDIATE rac1                FAILED OVER
ora.scan1.vip
     1      ONLINE  ONLINE      rac1
```

再看一下 Oracle 18c 中的 CRS 资源，代码如下。与上面的代码进行对比就可以发现多了一些资源，下面的代码中加粗显示了这些新增的资源。

```
[grid@RAC1 ~]$ crsctl stat res -t
--------------------------------------------------------------------
Name   Target  State      Server                  State details
--------------------------------------------------------------------
Local Resources
--------------------------------------------------------------------
ora.ASMNET1LSNR_ASM.lsnr
            ONLINE  ONLINE      rac1                  STABLE
ora.DATA.dg
            ONLINE  ONLINE      rac1                  STABLE
ora.LISTENER.lsnr
            ONLINE  ONLINE      rac1                  STABLE
ora.OCR_MGMT.GHCHKPT.advm
            OFFLINE OFFLINE     rac1                  STABLE
```

```
ora.OCR_MGMT.dg
               ONLINE  ONLINE       rac1                     STABLE
ora.chad
               OFFLINE OFFLINE      rac1                     STABLE
ora.helper
               OFFLINE OFFLINE      rac1                     IDLE,STABLE
ora.net1.network
               ONLINE  ONLINE       rac1                     STABLE
ora.ocr_mgmt.ghchkpt.acfs
               OFFLINE OFFLINE      rac1                     STABLE
ora.ons
               ONLINE  ONLINE       rac1                     STABLE
ora.proxy_advm
               ONLINE  ONLINE       rac1                     STABLE
--------------------------------------------------------------------------------
Cluster Resources
--------------------------------------------------------------------------------
ora.LISTENER_SCAN1.lsnr
      1        ONLINE  ONLINE       rac1                     STABLE
ora.MGMTLSNR
      1        ONLINE  ONLINE       rac1
                                                 169.254.10.77 192.16
                                                         8.3.101,STABLE
ora.asm
      1        ONLINE  ONLINE       rac1                     STABLE
      2        ONLINE  OFFLINE                               STABLE
      3        OFFLINE OFFLINE                               STABLE
ora.cndba.db
      1        ONLINE  ONLINE       rac1
Open,HOME=/u01/app/o

racle/product/18.3.0
                                                         /db_1,STABLE
      2        ONLINE  OFFLINE                               Instance
Shutdown,ST
                                                         ABLE
ora.cvu
      1        ONLINE  ONLINE       rac1                     STABLE
ora.mgmtdb
      1        ONLINE  ONLINE       rac1                     Open,STABLE
ora.qosmserver
      1        ONLINE  ONLINE       rac1                     STABLE
ora.rac1.vip
```

```
     1        ONLINE  ONLINE      rac1                    STABLE
ora.rac2.vip
     1        ONLINE  INTERMEDIATE rac1                    FAILED
OVER,STABLE
ora.rhpserver
     1        OFFLINE OFFLINE                             STABLE
ora.scan1.vip
     1        ONLINE  ONLINE      rac1                    STABLE
--------------------------------------------------------------
```

下面对新增的 CRS 资源进行简单的说明。

- ora.ASMNET1LSNR_ASM.lsnr：集群 ASM 的监听。
- ora.proxy_advm：这里的 ADVM 是 ASM Dynamic Volume Manager 的缩写，如果使用了 ASM 动态卷管理器，那么就会有这个资源。
- ora.MGMTLSNR：MGMT 数据库的监听。
- ora.mgmtdb：MGMT 数据库，在后面的章节中会介绍。
- ora.qosmserver：Oracle 数据库 QoS（Quality of Service）的管理。QoS 从 Oracle 12.1 开始被引入，是一个自动化的、基于策略的产品，用于监视整个系统的工作负载请求。Oracle 数据库的 QoS 管理跨应用程序共享资源，并调整系统配置，使应用程序运行在业务需要的性能级别。
- ora.rhpserver：快速资源调配（Rapid Home Provisioning）是一种用于配置和维护 Oracle Home 软件生命周期的管理方法，它能维护数据库、集群和用户定义软件类型的标准操作环境。通过它还可以安装集群和补丁，升级 Grid Infrastructure 和数据库（Oracle 11.2 及以后版本）等。另外，还可以配置应用程序和中间件。

12.5　RAC 数据库的配置类型

在创建 RAC 数据库时，会有数据库类型和配置类型的选择，如图 12-2 所示。数据库类型选择 RAC，配置类型有以下两种。

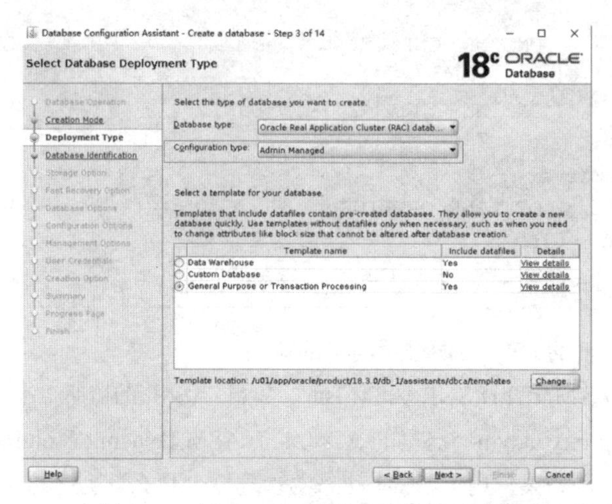

图 12-2 创建 RAC 数据库的设置对话框

- Admin Managed 是 Oracle 11gR2 之前安装 RAC 的配置类型，需要手动配置每个数据库实例在群集中的特定节点上运行。在安装时指定在某几个节点上运行数据库实例后，以后如果不做增减节点的操作，那么数据库实例始终运行在这几个节点上。
- Policy Managed 是以服务池（Server Pools）为基础的，先定义一些服务池，服务池中包含指定的节点，然后定义一些策略，根据这些策略，Oracle 自动决定让哪些数据库实例运行在池中的几台机器上。数据库实例名后缀、数据库实例个数、所运行的主机，这些都是通过策略决定的，不需要 DBA 事先设置好。

从上面的说明可以看出，服务池是 Policy Managed 数据库的基础。服务池可以将一组数据库、应用程序以逻辑的方式进行划分。根据定义好的策略（最大数量、最小数量、重要性）自动将数据库、应用程序等资源分配到不同的服务池中，这些操作不需要 DBA 来干预，Oracle 自动完成。

默认情况下，新安装的环境都会自动创建两个服务池：Generic 和 Free。在没有配置和创建其他服务池的情况下，所有服务器都会被分配到 Free 服务池中，再从 Free 服务池迁移到新创建的服务池中。

假设有一个 9 个节点的 RAC 集群，并且配置了 3 个服务器池（DBSP1、DBSP2、DBSP3），策略如下：

- DBSP1：最小数量 4，最大数量 6，重要性是 10；

- DBSP2：最小数量 2，最大数量 3，重要性是 6；
- DBSP3：最小数量 2，最大数量 3，重要性是 4。

那么，服务器池 DBSP1 将分配到节点 1~5，服务器池 DBSP2 将分配到节点 6 和节点 7，服务器池 DBSP3 将分配到节点 8 和节点 9。如果后面又新增了 2 个节点，那么其中一个节点要分配到 DBSP1 中，因为 DBSP1 的重要性最高，而另一个节点要分配给 DBSP2（因为 DBSP1 已经达到最大值了）。

从实际情况来看，目前生产环境大部分还是使用 Admin Managed 类型，可能和 RAC 节点的数量有一定关系，因为目前大部分的 RAC 环境数据节点的数量还是小于 5 个的，再加上业务类型，使用 Policy Managed 配置类型管理的实际意义并不大。另外，Policy Managed 也是在 Oracle 11gR2 这个版本中才推出的，在成为主流选择之前，可能还需要一点时间。

12.6　Hang 管理器概述

Hang 管理器用于自动检测和处理系统故障。它在 Oracle 11g R1 中被引入，在系统挂起时，自动将相关信息转存到 Trace 文件中以供 DBA 分析。在 Oracle 18c 中，Hang 管理器得到了加强，甚至可以自动尝试解决一些小问题，但 Hang 管理器无法解决死锁问题，因为它无法判断事务的重要性。

Hang 管理器有如下功能。

- 检测系统挂起，分析并验证挂起的原因，最后找出解决挂起的方法。
- 定期扫描所有进程并分析一直占用较多资源的部分进程。如果资源上没有任何等待，则 Hang 管理器会忽略该进程。
- 分析跨实例挂起，这里主要指数据库进程的持有者等待 Oracle ASM 实例的响应。
- 分析只读节点（Leaf）上的实例中运行的进程，并检查这些进程是否阻止 Hub 节点上的进程，如果是，则采取必要的措施。
- 查看并分析挂起持有者的 Oracle 数据库服务质量管理（Oracle Database Quality of Service Management）的相关设置。
- 关闭挂起持有者的进程，以便等待该资源的下一个进程可以继续使用并防止挂起。
- 将 ORA-32701 错误信息记录到告警日志中，以便 DBA 查看。

第 13 章

管理集群数据库和实例

在管理 RAC 之前需要搭建好 Oracle 18c 的 RAC 环境，因为官方手册中的搭建内容较全面，所以这里就不再赘述了。需要查看 Oracle 18c RAC 搭建内容的读者，可以浏览 CNDBA 社区的网站（https://www.cndba.cn），或者加入 CNDBA 社区的 QQ 交流群，在群里都有分享。

13.1 RAC 中的初始化参数

数据库安装成功后，Oracle 就会根据设置自动在相应的目录下创建 SPFILE 文件。Oracle 推荐使用 SPFILE 文件来启动数据库，因为通过 SPFILE 文件启动的数据库支持动态修改和动态参数。使用 RMAN 文件可以对 SPFIEL 文件进行备份。

RAC 中的所有数据库实例都公用一个 SPFILE 文件（也可以有各自的 SPFILE 文件，但不建议这么用，不同实例的不同参数值可能会造成数据库无法正常启动）。SPFILE 是二进制文件，不能直接修改，如果想修改数据库参数，则要使用 ALTER SYSTEM 文件语句，或者创建并修改 PFILE 文件后将其转换成 SPFILE 文件。

13.1.1 在 RAC 中设置 SPFILE 文件参数值

可以通过 PFILE 文件来修改 SPFILE 文件，或者直接通过命令来修改相应的参数值。SPFILE 被修改后，数据库重启后会检查 SPFILE 文件中的参数值，如果有参数不符合要求，则数据库无法打开。这时只能先创建 PFILE 文件，然后修改 PFILE 文件的相应参数，最后重新生成 SPFILE 文件。

PFILE 文件的参数前面支持两种前缀：一种是*号，表示对所有实例生效；

另一种是实例名，表示只对该实例生效。例如：

```
*.open_cursors=500
cndba1.open_cursors=1000
```

上面的参数设置语句表示除实例 cndba1 外的其他所有实例的 open_cursors 参数值都是 500，cndba1 实例的 open_cursors 参数值是 1000。

如果想修改参数，则可以使用如下方法，语法比较简单：

```
alter system set open_cursors=1500 sid='*' scope=spfile;
alter system reset open_cursors scope=spfile;
alter system reset open_cursors scope=spfile sid=' cndba1';
```

13.1.2　数据库被打开时搜索参数文件的位置

数据库被打开时会查找参数文件，如果查找不到，则数据库无法正常启动。不同的操作系统，参数文件存放的位置不同。在 RAC 中可以通过 srvctl config database 命令来修改参数文件的存放位置，代码如下：

```
[oracle@rac2 ~]$ srvctl config database -db cndba
Database unique name: cndba
Database name: cndba
Oracle home: /u01/app/oracle/product/18.3.0/db_1
Oracle user: oracle
Spfile: +DATA/CNDBA/PARAMETERFILE/spfile.278.986019415
Password file: +DATA/CNDBA/PASSWORD/pwdcndba.257.986017503
Domain:
Start options: open
Stop options: immediate
Database role: PRIMARY
Management policy: AUTOMATIC
Server pools:
Disk Groups: DATA
Mount point paths:
Services:
Type: RAC
Start concurrency:
Stop concurrency:
OSDBA group: dba
OSOPER group: oper
Database instances: cndba1,cndba2
Configured nodes: rac1,rac2
```

```
CSS critical: no
CPU count: 0
Memory target: 0
Maximum memory: 0
Default network number for database services:
Database is administrator managed
[oracle@rac2 ~]$
```

在 Linux 和 UNIX 平台上，参数文件的搜索顺序如下：

（1）$ORACLE_HOME/dbs/spfilesid.ora；

（2）$ORACLE_HOME/dbs/spfile.ora；

（3）$ORACLE_HOME/dbs/initsid.ora。

在 Windows 平台上，参数文件的搜索顺序如下：

（1）%ORACLE_HOME%\database\spfilesid.ora；

（2）%ORACLE_HOME%\database\spfile.ora；

（3）%ORACLE_HOME%\database\initsid.ora。

13.1.3 初始化参数在 RAC 中的使用

默认情况下，大部分参数值都被设置为默认值，并且在所有实例上都相同，但是有些参数需要特别注意。如表 13-1 所示是专门用于 Oracle RAC 数据库的初始化参数。

表 13-1 专门用于 Oracle RAC 数据库的初始化参数

参 数	说 明
CLUSTER_DATABASE	集群环境中的数据库，参数值一定是 TRUE
CLUSTER_DATABASE_INSTANCES	Oracle RAC 根据该参数来分配足够的内存资源，必须在所有实例上将其设置为相同的值。 • 对于 Policy Managed 数据库，默认将该参数设置为 16 • 对于 Admin Managed 数据库，Oracle 将该参数设置为配置的实例个数
CLUSTER_INTERCONNECTS	Oracle 不推荐设置该参数
DB_NAME	如果在参数文件中指定了该参数的值，那么就要保证所有实例的该参数值相同
DISPATCHERS	设置该参数是为了启用共享服务器配置，启用后允许多用户进程共享很少的服务器进程

续表

参　　数	说　　明
GCS_SERVER_PROCESSES	该参数为静态参数，用于指定 Oracle RAC 实例的全局缓存服务（GCS）的服务器进程的数量。GCS 进程管理 Oracle RAC 实例间通信的路由。GCS 服务器进程的默认数量是根据 CPU 数量计算得来的。如果有一个 CPU，那么就有一个 GCS 服务器进程；如果有 2~8 个 CPU，那么就有两个 GCS 服务器进程；如果 CPU 数量大于 8，那么 GCS 服务器进程的数量等于 CPU 的数量除以 4，然后取整。例如，如果有 10 个 CPU，那么用 10 除以 4 再取整，即有 2 个 GCS 进程。可以为不同的实例配置不同的值
INSTANCE_NAME	和 DB_NAME 相反，每个实例的 INSTANCE_NAME 的值都不相同
SERVICE_NAMES	服务名，可通过这些服务名连接数据库。在 tnsnames.ora 中配置的服务名需要与之对应
SPFILE	如果使用了 SPFILE 文件，那么该参数的取值是 SPFILE 文件的具体路径。建议使用 SPFILE 文件
THREAD	指定实例可以使用的 Redo 线程数量，所有实例都要保持一致

1．所有实例中参数值一定相同的参数

在 RAC 环境中，有些参数的值必须在所有节点上保持一致，否则可能会出现数据库性能问题，甚至导致数据库无法运行。所以，Oracle 极力推荐在 RAC 环境中将 SPFILE 文件存储到 ASM 中，所有数据库实例共享一个 SPFILE 文件，这样可以大大减小参数值不一致而出现数据库故障的概率。

以下参数必须设置为相同的参数值：

- COMPATIBLE；
- CLUSTER_DATABASE；
- CONTROL_FILES；
- DB_BLOCK_SIZE；
- DB_DOMAIN；
- DB_FILES；
- DB_NAME；
- DB_RECOVERY_FILE_DEST；
- DB_RECOVERY_FILE_DEST_SIZE；

- DB_UNIQUE_NAME；
- INSTANCE_TYPE （RDBMS or ASM）；
- PARALLEL_EXECUTION_MESSAGE_SIZE；
- REMOTE_LOGIN_PASSWORDFILE；
- UNDO_MANAGEMENT；
- DML_LOCKS；
- RESULT_CACHE_MAX_SIZE。

2. 所有实例中参数值一定不同的参数

在 RAC 环境中，以下参数在不同的实例中参数值一定不能相同。

- ORACLE_SID：通常由 DB_NAME 和 instance_number 组成。
- CLUSTER_INTERCONNECTS：用于配置集群中数据库之间的心跳网络，默认是禁用的。不推荐使用这种方式来配置心跳网络，使用 HAIP 或用网卡绑定更好。
- INSTANCE_NUMBER：不同实例的实例号不相同，Oracle 使用该参数来区分启动时的实例和 INSTANCE_NAME 参数，以将重做日志组分配给特定实例。查看参数 INSTANCE_NUMBER 的代码如下：

```
SQL> show parameter instance_number
NAME                 TYPE   VALUE
---------------- ----------- ---------------
instance_number   integer     1
```

- UNDO_TABLESPACE：为不同的实例分配不同的 UNDO 表空间。查看参数 UNDO_TABLESPACE 的代码如下：

```
SQL> show parameter UNDO_TABLESPACE
NAME                 TYPE   VALUE
--------------- ---------- ----------------
undo_tablespace   string  UNDOTBS1
```

- ROLLBACK_SEGMENETS：如果使用该参数，则 Oracle 建议使用 SPFILE 文件中的 SID 作为其唯一的值。注意，必须保证每个实例的 INSTANCE_NUMBER 的值不同。

13.2　启动/关闭数据库和实例

在 GI 运行正常的情况下，可以通过 SQLPLUS、SRVCTL 和 CRSCTL 命令来启动/关闭数据库和实例。这三个命令的功能非常多，这里不能进行逐一的讲解。在日常使用的过程中，若有命令记不清楚，可以查看命令的帮助文档。这三个命令的帮助文档非常详细，可以一层一层地查看。所谓的一层一层，就是每次增加一个命令选项，这样可以看到具体的使用方法。

下面演示如何一层一层地查看命令，代码如下：

```
[grid@rac1 ~]$ crsctl -h
Usage:
       crsctl add       - add a resource, type or other entity
       crsctl check     - check the state or operating status of a
service, resource, or other entity
       crsctl config    - display automatic startup configuration
       crsctl create    - display entity creation options
       crsctl debug     - display or modify debug state
…
[grid@rac1 ~]$ crsctl add -h
Usage:
  crsctl add
{resource|type|resourcegroup|resourcegrouptype|serverpool|policy}
<name> <options>
where
   name        Name of the CRS entity
   options     Options to be passed to the add command

   See individual CRS entity help for more details

  crsctl add crs administrator -u <user_name> [-f]
……
[grid@rac1 ~]$ crsctl add resource -h
Usage:
  crsctl add resource <resName> -type <typeName> [[-file <filePath>]
| [-attr "<specification>[,...]"]] [-f] [-i] [-group
<resourceGroupName>]
     <specification>:  {<attrName>=<value> |
<attrName>@<scope>=<value>}
       <scope>:  {@SERVERNAME (<server>) [@DEGREEID (<did>) ] |
                  @CARDINALITYID (<cid>) [@DEGREEID (<did>) ] }
where
```

```
    resName          Add named resource
    typeName         Resource type
    filePath         Attribute file
    attrName         Attribute name
    value            Attribute value
    server           Server name
    cid              Resource cardinality ID
    did              Resource degree ID
    -f               Force option
    -i               Fail if request cannot be processed immediately
    resourceGroupName  Resource group name
```

只要掌握了查看命令帮助文档的方法，命令就不需要刻意地记忆了。下面讲解最基本的数据库和实例的启动与关闭。

13.2.1 使用 SRVCTL 命令启动数据库和实例

1．启动整个集群数据库（所有实例及其启用的服务）

启动整个集群数据库，代码如下：

```
    srvctl start database -db db_unique_name [-startoption
start_options]
```

2．启动并挂载数据库

启动并挂载数据库，代码如下：

```
    srvctl start database -db cndba -startoption mount
```

3．启动集群中的指定实例

- 启动 admin Managed 数据库，代码如下：

```
    srvctl start instance -db db_unique_name -instance
instance_name_list [-startoption start_options]
    srvctl start instance -db cndba -instance cndba1,cndba2
```

如果是 Windows 环境，则实例名称要使用双引号，如“cndba1,cndba2”。

- 启动 policy Managed 数据库，代码如下：

```
    srvctl start instance -db db_unique_name -node node_name
[-startoption start_options]
    srvctl start instance -db cndba -node rac1
```

13.2.2　使用 SRVCTL 命令关闭数据库和实例

关闭 RAC 实例与关闭普通的单实例是一样的，但是以下情况需要特别注意。

- 在 RAC 环境中，关闭一个实例不会影响其他实例。
- 要完全关闭 RAC 数据库，需要关闭集群中所有已打开的数据库实例。
- 使用 NORMAL 或 IMMEDIATE 关闭实例后，不需要实例恢复。但是，在执行 SHUTDOWN ABORT 命令或实例异常终止后，需要进行实例恢复。仍在运行的实例会为异常关闭的实例提供实例恢复。如果没有其他实例正在运行，则第一个打开的数据库实例将为其他需要进行恢复的实例提供实例恢复。
- 使用带有 LOCAL 选项的 SHUTDOWN TRANSACTIONAL 命令对于关闭指定的 Oracle RAC 数据库实例很有用。其他实例上的事务不会阻塞此操作。如果省略 LOCAL 选项，则此操作会等待其他实例上的事务提交或回滚，才能完成实例的关闭。

1．关闭整个集群数据库（所有实例及其启用的服务）

关闭整个集群数据库，代码如下：

```
srvctl stop database -db db_unique_name [-stopoption stop_options]
```

2．立即关闭数据库

立即关闭数据库，代码如下：

```
srvctl stop database -db cndba - stopoption immediate
```

3．关闭集群中的指定实例

- 关闭 Admin Managed 数据库，代码如下：

```
    srvctl stop instance -db db_unique_name -instance
instance_name_list [-stopoption start_options]
    srvctl stop instance -db cndba -instance cndba1,cndba2 -stopoption
immediate
```

- 关闭 Policy Managed 数据库，代码如下：

```
    srvctl stop instance -db db_unique_name -node node_name
[-stopoption start_options]
    srvctl stop instance -db cndba -node rac1 -stopoption immediate
```

13.2.3 使用 CRSCTL 命令启动/关闭所有数据库和实例

如果想使用 CRSCTL 命令来关闭数据库和实例，只能通过关闭整个集群来实现。该方法相当于使用 SHUTDOWN ABORT 方式关闭数据库，在数据库再次被打开后需要进行实例恢复。如果只是关闭数据库和实例，则这种方式不推荐使用。执行以下命令需要用 root 账户。

关闭 CRS：crsctl stop crs

启动 CRS：crsctl start crs

关闭整个集群：crsctl stop cluster -all

启动整个集群：crsctl start cluster -all

13.2.4 使用 SQLPLUS 命令启动/关闭实例

使用 SQLPLUS 命令启动/关闭数据库实例和使用 SQLPLUS 命令启动/关闭单实例没有任何区别，需要注意关闭之前要查看当前的 SID 是否正确。

- 以 immediate 方式关闭实例，代码如下：

```
sqlplus / as sysdba
shutdown immediate
```

- 打开数据库实例，代码如下：

```
sqlplus / as sysdba
startup
```

也可以使用 SQLPLUS 命令关闭 ASM 实例，但是 Oracle 不推荐这么做。ASM 实例由 GI 来维护，如果需要手动进行配置和重启操作，建议使用 SRVCTL 命令。另外需要注意的是，如果通过 TNS 别名连接数据库实例，则一定要确认当前连接的实例正确，因为在 TNS 中，可能是通过 SCAN IP 随机连接实例的，可能会连接到别的实例上。

13.3 Oracle 日志结构

从 Oracle 12c 开始，日志的存储位置和命名方式发生了比较大的变化，而数据库的告警日志的存储位置和命名方式没有变化，还是存储在 ORACLE_BASE/diag/rdbms/db_name/SID/目录下。日志的主要变化体现在 RAC

集群日志上，特别是存储位置发生了比较大的变化。从 Oracle 12.1.0.2 开始，集群日志使用了新的命名规则，所有与集群相关的日志文件都以.trc 为扩展名，除了集群的告警日志（Oracle 12c 中是 alert.log，Oracle 12c 之前是 alert<hostname>.log）。

下面看一下 ADR 中不同产品/组件的子目录结构，可以看到集群中与不同组件相关的日志存储在不同的目录下，如图 13-1 所示。

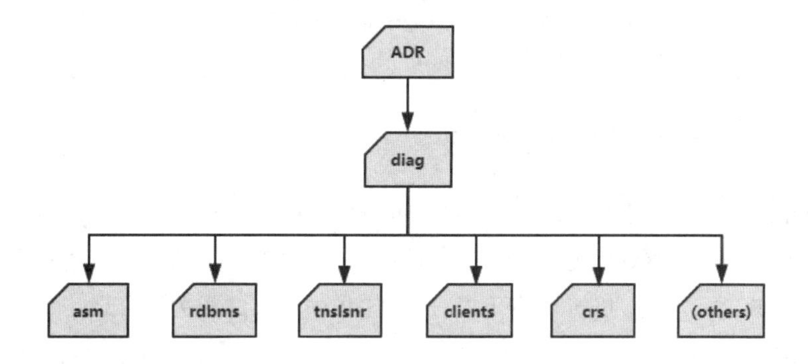

图 13-1　ADR 中不同产品/组件的子目录结构

ADR 中数据库实例的目录结构如图 13-2 所示。子目录中包含的内容如表 13-2 所示。

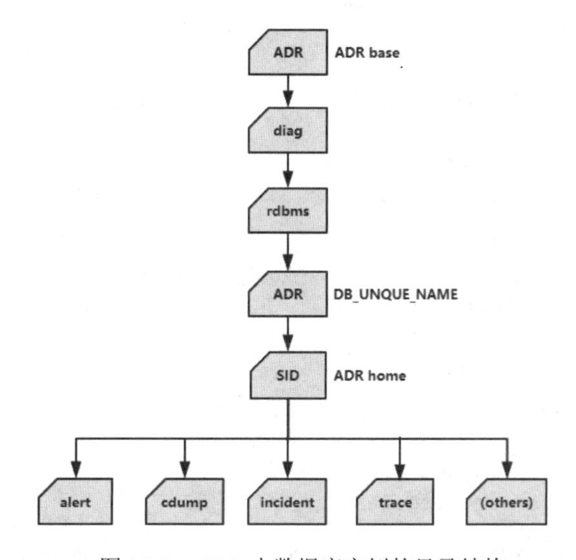

图 13-2　ADR 中数据库实例的目录结构

表 13-2　ADR 中数据库实例的子目录名称及内容

子目录名称	内　容
alert	XML 格式的告警日志
cdump	内核文件
incident	有多个子目录，每个子目录都是以特定事件来命名的，每个子目录包含仅与该事件有关的转储内容
trace	后台进程，服务进程的 Trace 文件、SQL Trace 文件和文本格式的告警日志
（others）	ADR home 下的其他子目录，用于存储事件包、健康监控报告、告警日志以外的日志，如 DDL 日志和调试日志

13.3.1　Oracle 11g 日志结构

Oracle 从 11g 开始引入了 ADR（Automatic Diagnostic Repository），数据库、自动存储管理（ASM）、CRS 和其他 Oracle 产品（组件）将所有诊断数据都存储在 ADR 中。通过 adrci 命令可以查看 ADR 中与日志相关的目录。

Grid 账户：

```
[grid@rac1 ~]$ adrci
ADRCI: Release 11.2.0.4.0 - Production on Mon Sep 10 10:05:21 2018
Copyright (c) 1982, 2011, Oracle and/or its affiliates.  All rights reserved.
ADR base = "/u01/gridbase"
adrci> show homes
ADR Homes:
diag/asm/+asm/+ASM1
diag/tnslsnr/rac1/listener
```

Oracle 账户：

```
[oracle@rac1 ~]$ adrci
ADRCI: Release 11.2.0.4.0 - Production on Mon Sep 10 11:10:53 2018
Copyright (c) 1982, 2011, Oracle and/or its affiliates.  All rights reserved.
ADR base = "/u01/oracle"
adrci> show homes
ADR Homes:
diag/rdbms/cndba/cndba1
diag/clients/user_oracle/host_1874443374_80
```

Oracle 11g 的数据库、ASM、与集群相关的日志的存储路径分别如下。

- 数据库告警日志：$ORACLE_BASE/diag/rdbms/db_name/SID/trace/alert_ <SID>.log。
- ASM 日志：GRID_BASE/diag/asm/+asm/ASM_SID/trace/alert_<ASM SID>.log。
- 集群告警日志：$ORACLE_HOME/log/<HOSTNAME>/alert <hostname>.log。
- HAS：HAS 的日志位于集群日志路径下的单独目录中。
- OHASD：$ORACLE_HOME/log/<HOSTNAME>/ohasd/ohasd.log。
- CSS：$ORACLE_HOME/log/<HOSTNAME>/cssd/cssd.log。

13.3.2　Oracle 18c 日志结构

Oracle 从 12c 开始，集群日志的存储位置和命名方式都发生了比较大的变化，日志完全由 ADR 接管，与集群相关的日志的存储路径可以从下面的命令中查看到。不同账户运行 adrci 命令查看到的日志信息是不一样的。要查看与集群相关的日志信息，需要用 Grid 账户运行 adrci 命令；要查看与数据库相关的日志信息，需要用 Oracle 账户运行 adrci 命令。

用 Grid 账户查看 ADR 中与集群相关的日志信息，代码如下：

```
[grid@rac1 ~]$ adrci
ADRCI: Release 18.0.0.0.0 - Production on Mon Sep 10 10:54:09 2018
Copyright (c) 1982, 2018, Oracle and/or its affiliates.  All rights reserved.

ADR base = "/u01/app/grid"
adrci> show homes
ADR Homes:
diag/rdbms/_mgmtdb/-MGMTDB  --MGMT 数据库的告警日志
diag/asm/+asm/+ASM1   --ASM 日志
diag/crs/rac1/crs  --集群日志
diag/clients/user_grid/host_1874443374_110
diag/tnslsnr/rac1/asmnet1lsnr_asm  --ASM 监听日志
diag/tnslsnr/rac1/listener_scan1   --scan 监听日志
diag/tnslsnr/rac1/listener   --监听日志
diag/tnslsnr/rac1/mgmtlsnr  --MGMT 监听日志
diag/asmtool/user_grid/host_1874443374_110
diag/asmcmd/user_grid/rac1
diag/asmcmd/user_root/rac1
```

```
diag/kfod/rac1/kfod
```

与 Oracle 11g 不同的是，Oracle 18c 中所有与集群资源相关的日志都存储在集群日志目录中，而且全部存放在 trace 目录中（ADR_BASE/diag/crs/rac1/crs），不再存放在单独的目录中。

用 Oracle 账户查看与数据库相关的日志，代码如下：

```
[oracle@rac1 ~]$ adrci
ADRCI: Release 18.0.0.0.0 - Production on Mon Sep 10 11:00:17 2018
Copyright (c) 1982, 2018, Oracle and/or its affiliates.  All rights
reserved.

ADR base = "/u01/app/oracle"
adrci> show homes
ADR Homes:
diag/rdbms/cndba/cndba1 --数据库告警日志
diag/clients/user_oracle/host_1874443374_110
```

也可以通过 V$DIAG_INFO 视图查看数据库实例所有重要的 ADR 存储位置，代码如下：

```
SQL> col value for a50
SQL> set line 200
SQL> SELECT * FROM V$DIAG_INFO;
   INST_ID NAME              VALUE
CON_ID
---------------------------------------------------- ----------
   1 Diag Enabled   TRUE                                      0
   1 ADR Base        /u01/app/oracle                          0
   1 ADR Home        /u01/app/oracle/diag/rdbms/cndba/cndba1   0
   1 Diag Trace   /u01/app/oracle/diag/rdbms/cndba/cndba1/trace 0
   1 Diag Alert
/u01/app/oracle/diag/rdbms/cndba/cndba1/alert            0
   1 Diag Incident
/u01/app/oracle/diag/rdbms/cndba/cndba1/incident          0
   1 Diag Cdump
/u01/app/oracle/diag/rdbms/cndba/cndba1/cdump            0
   1 Health Monitor
/u01/app/oracle/diag/rdbms/cndba/cndba1/hm             0
   1 Default Trace File
/u01/app/oracle/diag/rdbms/cndba/cndba1/trace/cndb a1_ora_13395.trc 0
   1 Active Problem Count 0                                   0
   1 Active Incident Count 0                                  0
```

使用 adrci 命令中的 show homes 子命令就可以查看数据库和集群中各种日志的存储位置。下面列举几个常见的日志存储位置。

- 数据库告警日志：Oracle 18c 的数据库告警日志的存储位置和命名方式与之前版本的数据库告警日志的储存位置和命名方式没有区别。存储位置为$ORACLE_BASE/diag /rdbms/db_name/SID/ trace/alert_<SID>.log。
- ASM 日志：Oracle 18c 的 ASM 日志的存储位置与之前版本的 ASM 日志的存储位置一样。存储位置均为$GRID_BASE/diag/asm/+asm/ASM_SID/trace/alert_<ASM SID>.log。
- 集群告警日志：Oracle 18c 集群告警日志的存储路径和命名方式都发生了变化。日志存储位置为$GRID_BASE/diag/crs/rac1/crs/trace/alert.log。
- HAS：从 Oracle 12c 开始，集群中各个进程的日志、trace 文件和集群告警日志存储在同一个目录下。存储位置为$GRID_BASE/diag/ crs/rac1/crs/trace/ohasd.trc。可以看到，扩展名从.log 改为了.trc。只有集群的告警日志的扩展名是.log，其他 CRS 中的资源或进程的日志扩展名都是.trc。
- CSS：存储位置为 $GRID_BASE/diag/crs/rac1/crs/trace/cssd.trc。

从上面的说明可以看出，Oracle 18c 中发生最大变化的是集群日志，了解日志位置的变化后就可以快速地查看日志并进行故障分析了。

13.4　RAC 中的 Kill 会话

可以使用 ALTER SYSTEM KILL SESSION 命令终止指定实例上的会话。当一个会话被终止后，该会话上的所有事务将被回滚，并且该会话持有的资源（占有的锁、内存）也会被立即释放。在 RAC 中，终止会话的具体操作步骤如下。

（1）通过 GV$SESSION 视图查找 INST_ID、SID、SERIAL#，代码如下：

```
SQL> SELECT SID, SERIAL#, INST_ID FROM GV$SESSION WHERE
USERNAME='LEI';
  SID   SERIAL#  INST_ID
---------- ----------- --------------
  80      4       2
```

（2）终止实例 2 上的会话，代码如下：

```
SQL> ALTER SYSTEM KILL SESSION '80, 4, @2';
System altered.
```

（3）操作中可能会遇到 ORA-00031 错误，代码如下：

```
SQL> ALTER SYSTEM KILL SESSION '80, 4, @2';
alter system kill session '80, 4, @2'
*
ERROR at line 1:
ORA-00031: session marked for kill
SQL>
```

错误 ORA-00031 表示在终止该会话之前有些活动的事务需要完成，这里先将该会话标记为终止。在真正终止之前，资源不会被释放。也可以使用 IMMEDIATE 命令立即终止会话，而不需要等待其他事务完成，代码如下：

```
ALTER SYSTEM KILL SESSION '80, 4, @2' IMMEDIATE;
```

13.5 管理 OCR 和 OLR

Oracle 集群软件中有三个重要的组件：OCR（Oracle Cluster Registry）、OLR（Oracle Local Registry）和 Voting File。

- OCR 存储 Oracle 集群和 RAC 数据库的配置信息。
- OLR 存储在集群的每个节点上，并管理每个节点的 Oracle 集群配置信息。
- Voting File 存储有关节点成员的信息。集群中的所有节点均可以访问每个 Voting File，从而使每个节点都成为集群中的一员

这里要注意一点，从 Oracle 12c 开始已经不支持裸设备或块设备，所以 Oracle 12c 之前的 RAC 升级到 Oracle 12c 及以后版本，需要在升级前将 OCR 和 Voting File 迁移到 Oracle ASM 存储上或共享文件系统上。可以使用 OCRCONFIG、OCRDUMP 和 OCRCHECK 命令管理 OCR 和 OLR。

OCR 包含集群中所有与 Oracle 相关的资源信息。OLR 是一个类似于存储在本地的 OCR，只包含该节点指定的信息。OCR 包含有关 Oracle 集群软件的可管理性信息，如各种服务之间的依赖关系，Oracle 高可用性服务会使用这些信息。OLR 存储在集群的每个节点上，默认位置是 Grid_home/cdata/host_name.olr。示例如下：

```
[root@rac1 install]# ll /u01/app/18.1.0/grid/cdata/rac1.olr
-rw------- 1 root oinstall 503484416 Jul 26 16:04
/u01/app/18.1.0/grid/cdata/rac1.olr
```

13.5.1　迁移 OCR 到 ASM

将 OCR 从其他设备迁移到 ASM 中的操作步骤如下。

（1）创建磁盘组来存储 OCR，代码如下：

```
CREATE DISKGROUP OCR_VOTING NORMAL REDUNDANCY
   '/dev/asmdiskh' NAME disk1,
   '/dev/asmdiski' NAME disk2,
   '/dev/asmdiskj' NAME disk3
 ATTRIBUTE 'au_size'='4M',
   'compatible.asm' = '18.0',
   'compatible.rdbms' = '18.0',
   'compatible.advm' = '18.0';
```

（2）将磁盘组添加到 OCR 配置中，最多可以同时添加五个不同的 OCR 存储位置，代码如下：

```
ocrconfig -add +OCR_VOTING
```

（3）删除原来的 OCR 存储位置，代码如下：

```
ocrconfig -delete /dev/raw/raw1
ocrconfig -delete /dev/raw/raw2
ocrconfig -delete /dev/raw/raw3
```

13.5.2　添加和删除一个 OCR 存储

Oracle 最多支持五个不同的 OCR 存储位置，可以使用 root 账户执行 ocrconfig 命令来对 OCR 的存储位置进行添加和删除。因为 OCR 对 RAC 来说至关重要，所以 Oracle 推荐使用以下策略来保护 OCR。

- 如果 OCR 存储没有做冗余设置（如 RAID），那么强烈建议将 OCR 做镜像存储（ASM 磁盘组）。
- 如果 OCR 存储在外部冗余的磁盘组中，那么至少要将 OCR 存储到两个独立的磁盘组中。

1. 添加一个 OCR 存储

添加一个 OCR 存储的操作步骤如下。

（1）添加 OCR 存储位置，代码如下：

```
[root@rac1 ~]# /u01/app/18.1.0/grid/bin/ocrconfig -add +data
```

（2）验证 OCR，代码如下：

```
 [root@rac1 ~]# /u01/app/18.1.0/grid/bin/ocrcheck
Status of Oracle Cluster Registry is as follows :
       Version                  :        4
       Total space （kbytes）     :    409568
       Used space （kbytes）      :      3108
       Available space （kbytes） :    406460
       ID                    : 715973296
       Device/File Name         :  +OCR_MGMT
                             Device/File integrity check
succeeded
       Device/File Name         :  +DATA
                             Device/File integrity check
succeeded
                             Device/File not configured
                             Device/File not configured
                             Device/File not configured
       Cluster registry integrity check succeeded
       Logical corruption check succeeded
```

可以看到，+DATA 磁盘组已经是 OCR 存储的一部分了。

2. 删除一个 OCR 存储

删除一个 OCR 存储的操作步骤如下。

（1）删除 OCR 存储位置，代码如下：

```
[root@rac1 ~]# /u01/app/18.1.0/grid/bin/ocrconfig -delete +data
```

（2）验证 OCR，代码如下：

```
[root@rac1 ~]# /u01/app/18.1.0/grid/bin/ocrcheck
Status of Oracle Cluster Registry is as follows :
       Version                  :   4
       Total space （kbytes）     :    409568
       Used space （kbytes）      :      3108
       Available space （kbytes） :    406460
```

```
        ID                    : 715973296
        Device/File Name         : +OCR_MGMT
                                 Device/File integrity check
succeeded

                                 Device/File not configured
                                 Device/File not configured
                                 Device/File not configured
                                 Device/File not configured
        Cluster registry integrity check succeeded
    Logical corruption check succeeded
```

3. 替换 OCR 存储

当某个磁盘组中的磁盘出现问题，其冗余级别已经不能保证数据安全时，就需要给该磁盘组添加新的磁盘，或者将 OCR 存储到其他磁盘组上。

替换 OCR 存储的操作步骤如下。

（1）检查当前 OCR 存储，代码如下：

```
    [root@rac1 ~]# /u01/app/18.1.0/grid/bin/ocrcheck
    Status of Oracle Cluster Registry is as follows :
        Version                  :        4
        Total space （kbytes）     :    409568
        Used space （kbytes）      :      3108
        Available space （kbytes） :    406460
        ID                    : 715973296
        Device/File Name         : +OCR_MGMT
                                 Device/File integrity check
succeeded

                                 Device/File not configured
                                 Device/File not configured
                                 Device/File not configured
                                 Device/File not configured
        Cluster registry integrity check succeeded
        Logical corruption check succeeded
```

（2）添加新的 OCR 存储，代码如下：

```
    [root@rac1 ~]# /u01/app/18.1.0/grid/bin/ocrconfig -add +data
```

这里可以通过添加并删除磁盘组的方式来实现替换，或者直接使用 replace 命令，代码如下：

```
    [root@rac1 ~]# /u01/app/18.1.0/grid/bin/ocrconfig -replace
+ocr_mgmt -replacement +data
```

```
PROT-29: The Oracle Cluster Registry location is already configured
```

（3）查看 OCR，代码如下：

```
 [root@rac1 ~]# /u01/app/18.1.0/grid/bin/ocrcheck
Status of Oracle Cluster Registry is as follows :
       Version                  :        4
       Total space （kbytes）     :    409568
       Used space （kbytes）      :     3108
       Available space （kbytes） :    406460
       ID                      : 715973296
       Device/File Name         : + OCR_MGMT
                                Device/File integrity check
succeeded
       Device/File Name         : + DATA
                                Device/File integrity check
succeeded
                                Device/File not configured
                                Device/File not configured
                                Device/File not configured
       Cluster registry integrity check succeeded
       Logical corruption check succeeded
```

（4）删除旧的 OCR 存储，代码如下：

```
[root@rac1 ~]# /u01/app/18.1.0/grid/bin/ocrconfig -delete
+ocr_mgmt
```

（5）再次检查 OCR，代码如下：

```
[root@rac1 ~]# /u01/app/18.1.0/grid/bin/ocrcheck
Status of Oracle Cluster Registry is as follows :
       Version                  :        4
       Total space （kbytes）     :    409568
       Used space （kbytes）      :     3180
       Available space （kbytes） :    406388
       ID                      : 715973296
       Device/File Name         : +DATA
                                Device/File integrity check
succeeded
                                Device/File not configured
                                Device/File not configured
                                Device/File not configured
                                Device/File not configured
       Cluster registry integrity check succeeded
```

```
Logical corruption check succeeded
```

4. 修复 OCR

如果由于断电或其他问题造成 OCR 损坏，那么可以使用 ocrconfig -repair 命令修复 OCR。在下面的代码中，OCR 存储在+DATA 和+OCR_MGMT 磁盘组上，利用+DATA 磁盘组来修复+OCR_MGMT 磁盘组。

```
ocrconfig -repair -replace +OCR_MGM -replacement +DATA
```

13.5.3 备份 OCR

OCR 备份分为手动备份和自动备份。

- 自动备份：Oracle 集群每四个小时自动创建一次 OCR 备份。在任何时候，Oracle 数据库始终保留五份 OCR，即最近的三份备份、前一天的备份和上一周的备份。自动备份策略无法修改。
- 手动备份：在 Oracle 集群运行的节点上执行 ocrconfig -manualbackup 命令，可以手动备份 OCR（OLR 只能进行手动备份）。另外，在运行 root.sh 脚本时会自动备份 OLR。

1. 手动备份 OCR

手动备份 OCR 的操作步骤如下。

（1）从 Oracle 12.2 开始，OCR 自动备份到磁盘组中。在 Oracle 12.2 之前，OCR 被备份到本地的文件系统中。查看 OCR 备份，代码如下：

```
[grid@rac1 ~]$ ocrconfig -showbackup [auto|manual]
rac1     2018/07/26 11:12:55
+OCR_MGMT:/rac/OCRBACKUP/backup00.ocr.286.982494773     0
rac1     2018/07/26 07:12:52
+OCR_MGMT:/rac/OCRBACKUP/backup01.ocr.285.982480369     0
rac1     2018/07/26 03:12:48
+OCR_MGMT:/rac/OCRBACKUP/backup02.ocr.282.982465965     0
rac1     2018/07/25 03:12:25
+OCR_MGMT:/rac/OCRBACKUP/day.ocr.287.982379545     0
rac1     2018/07/24 19:12:17
+OCR_MGMT:/rac/OCRBACKUP/week.ocr.284.982350739     0
PROT-25: Manual backups for the Oracle Cluster Registry are not
available
```

（2）手动备份 OCR，代码如下：

```
[root@rac1 ~]# /u01/app/18.1.0/grid/bin/ocrconfig -manualbackup
rac1 2018/07/26 17:57:17
+OCR_MGMT:/rac/OCRBACKUP/backup_20180726_175717.ocr.288.982519037
0
```

（3）再次查看 OCR 备份，代码如下：

```
[grid@rac1 ~]$ ocrconfig -showbackup
rac1     2018/07/26 11:12:55
+OCR_MGMT:/rac/OCRBACKUP/backup00.ocr.286.982494773     0
rac1     2018/07/26 07:12:52
+OCR_MGMT:/rac/OCRBACKUP/backup01.ocr.285.982480369     0
rac1     2018/07/26 03:12:48
+OCR_MGMT:/rac/OCRBACKUP/backup02.ocr.282.982465965     0
rac1     2018/07/25 03:12:25
+OCR_MGMT:/rac/OCRBACKUP/day.ocr.287.982379545     0
rac1     2018/07/24 19:12:17
+OCR_MGMT:/rac/OCRBACKUP/week.ocr.284.982350739     0
rac1   2018/07/26 17:57:17
+OCR_MGMT:/rac/OCRBACKUP/backup_20180726_175717.ocr.288.982519037 0
```

2．修改 OCR 备份位置

OCR 备份是存放在 OCR 磁盘组里的，这个路径可以修改，代码如下：

```
[root@rac1 ~]# /u01/app/18.1.0/grid/bin/ocrconfig -backuploc
+data
```

再次备份，可以看到已经备份到+DATA 磁盘组中，代码如下：

```
[root@rac1 ~]# /u01/app/18.1.0/grid/bin/ocrconfig -manualbackup
rac1     2018/07/26 18:02:07
+data:/rac/OCRBACKUP/backup_20180726_180207.ocr.324.982519327     0
rac1     2018/07/26 17:57:17
+OCR_MGMT:/rac/OCRBACKUP/backup_20180726_175717.ocr.288.982519037
0
```

13.5.4 利用 OCR 备份恢复 OCR

在 OCR 备份有效的情况下，可以通过 OCR 备份来恢复 OCR。如果 OCR 备份都失效，那么可以执行 root.sh 命令重建 OCR，这里不讲解重建 OCR，只讲解通过 OCR 备份恢复 OCR。

1．模拟 OCR 损坏

模拟 OCR 损坏的操作步骤如下。

（1）检查 OCR，代码如下：

```
[root@rac1 ~]# /u01/app/18.1.0/grid/bin/ocrcheck
Status of Oracle Cluster Registry is as follows :
       Version                  :        4
       Total space （kbytes）     :    409568
       Used space （kbytes）      :     3180
       Available space （kbytes） :    406388
       ID                     : 715973296
       Device/File Name         :  +OCR_TEST
                             Device/File integrity check
succeeded

                             Device/File not configured
                             Device/File not configured
                             Device/File not configured
                             Device/File not configured
       Cluster registry integrity check succeeded
       Logical corruption check succeeded
```

（2）手动备份 OCR，代码如下：

```
    [root@rac1 ~]# /u01/app/18.1.0/grid/bin/ocrconfig -manualbackup
    rac1  2018/07/27 09:41:21
+ocr_mgmt:/rac/OCRBACKUP/backup_20180727_094121.ocr.290.982575681
0
    rac1  2018/07/26 18:02:07
+data:/rac/OCRBACKUP/backup_20180726_180207.ocr.324.982519327    0
    rac1 2018/07/26 17:57:17
+OCR_MGMT:/rac/OCRBACKUP/backup_20180726_175717.ocr.288.982519037
0
```

（3）使用 dd 命令破坏磁盘。

首先查看磁盘组中有哪些磁盘，代码如下：

```
ASMCMD> lsdsk -t -G ocr_test
Create_Date  Mount_Date  Repair_Timer  Path
27-JUL-18   27-JUL-18   0           /dev/asmdiskh
```

在/dev/asmdiskh 磁盘中使用 dd 命令破坏磁盘头的数据，代码如下：

```
[root@rac1 ~]# dd if=/dev/zero of=/dev/asmdiskh bs=4096 count=1
1+0 records in
```

```
1+0 records out
4096 bytes (4.1 kB) copied, 9.5908e-05 s, 42.7 MB/s
```

2. 恢复 OCR

恢复 OCR 的操作步骤如下。

以下所有操作除特殊说明外，都用 root 账户运行。

（1）关闭 CRS（所有节点都执行），代码如下：

```
[root@rac1 bin] ./crsctl stop crs
```

如果无法正常关闭 CRS，则加参数-f 强行关闭。

（2）尝试正常启动 CRS，代码如下：

```
    2018-07-27 09:53:15.046 [OCRCHECK (9963) ]CRS-1013: The OCR
location in an ASM disk group is inaccessible. Details in
/u01/app/grid/diag/crs/rac1/crs/trace/ocrcheck_9963.trc.
    2018-07-27 09:53:15.144 [OCRCHECK (9963) ]CRS-1011: OCR cannot
determine that the OCR content contains the latest updates. Details
in .
```

从上述代码中可以看到无法正常启动 CRS。

（3）强制关闭 CRS，代码如下：

```
[root@rac1 bin]# ./crsctl stop crs -f
```

（4）以独占模式启用 CRS，代码如下：

```
[root@rac1 bin] ./crsctl start crs -excl -nocrs
```

-nocrs 表示 CRSD 进程和 OCR 都不启动。启动过程中的任何错误都不要管。

（5）检查 CRSD 是否启动，代码如下：

```
[root@rac1 bin]# ./crsctl status resource ora.crsd -init
NAME=ora.crsd
TYPE=ora.crs.type
TARGET=OFFLINE
STATE=OFFLINE
```

如果 CRSD 启动了，那么可以通过命令关闭它，代码如下：

```
[root@rac1 bin] ./crsctl stop resource ora.crsd -init
```

（6）在恢复 OCR 之前，需要创建一个和之前磁盘组名称一样的磁盘组并挂载，代码如下：

```
[grid@rac1 ~]$ sqlplus / as sysasm
SQL> CREATE DISKGROUP OCR_TEST EXTERNAL REDUNDANCY DISK
'/dev/asmdiskh' ATTRIBUTE 'COMPATIBLE.ASM'= '12.2.0';
Diskgroup created.
```

（7）使用 OCR 备份恢复 OCR。

首先查看 OCR 备份的位置，代码如下：

```
[root@rac1 bin]# ./ocrconfig -showbackup
PROT-26: Oracle Cluster Registry backup locations were retrieved
from a local copy
rac1     2018/07/27 09:18:40
+ocr_mgmt:/rac/OCRBACKUP/backup00.ocr.286.982574317     0
rac1     2018/07/27 05:18:34
+ocr_mgmt:/rac/OCRBACKUP/backup01.ocr.285.982559911     0
rac1     2018/07/27 01:18:30
+ocr_mgmt:/rac/OCRBACKUP/backup02.ocr.282.982545507     0
rac1     2018/07/26 03:12:48
+OCR_MGMT:/rac/OCRBACKUP/day.ocr.283.982465969     0
rac1     2018/07/24 19:12:17
+OCR_MGMT:/rac/OCRBACKUP/week.ocr.284.982350739     0
rac1     2018/07/26 18:02:07
+data:/rac/OCRBACKUP/backup_20180726_180207.ocr.324.982519327     0
rac1     2018/07/26 17:57:17
+OCR_MGMT:/rac/OCRBACKUP/backup_20180726_175717.ocr.288.982519037
0
```

使用 ocrconfig -showbackup 命令查看 OCR 备份，并用它来恢复 OCR，代码如下：

```
[root@rac1 bin]# ./ocrconfig -restore
+ocr_mgmt:/rac/OCRBACKUP/backup00.ocr.286.982574317
```

（8）验证 OCR 的完整性，代码如下：

```
[root@rac1 bin]# ./ocrcheck
Status of Oracle Cluster Registry is as follows :
        Version                  :         4
        Total space (kbytes)     :    409568
        Used space (kbytes)      :      2996
        Available space (kbytes) :    406572
        ID                    :  715973296
        Device/File Name         :  +OCR_TEST
                              Device/File integrity check
```

```
succeeded
                              Device/File not configured
                              Device/File not configured
                              Device/File not configured
                              Device/File not configured
        Cluster registry integrity check succeeded
        Logical corruption check succeeded
```

（9）关闭 CRS，代码如下：

```
[root@rac1 bin] ./crsctl stop crs
```

如果无法正常关闭 CRS，则加参数-f 强行关闭。

（10）启用 CRS（所有节点），代码如下：

```
[root@rac1 bin] ./crsctl start crs
```

（11）查看集群，代码如下：

```
[root@rac1 bin]# ./crsctl stat res -t
--------------------------------------------------------------
------------------
Name           Target  State        Server              State details
--------------------------------------------------------
Local Resources
--------------------------------------------------------------
ora.ASMNET1LSNR_ASM.lsnr
               ONLINE  ONLINE       rac1                STABLE
               ONLINE  ONLINE       rac2                STABLE
ora.DATA.dg
               ONLINE  ONLINE       rac1                STABLE
               ONLINE  ONLINE       rac2                STABLE
ora.LISTENER.lsnr
               ONLINE  ONLINE       rac1                STABLE
               ONLINE  ONLINE       rac2                STABLE
ora.OCR_MGMT.dg
               ONLINE  ONLINE       rac1                STABLE
               ONLINE  ONLINE       rac2                STABLE
ora.OCR_TEST.dg
               OFFLINE OFFLINE      rac1                STABLE
               ONLINE  ONLINE       rac2                STABLE
ora.chad
               ONLINE  ONLINE       rac1                STABLE
               ONLINE  ONLINE       rac2                STABLE
ora.net1.network
```

```
                ONLINE  ONLINE       rac1                     STABLE
                ONLINE  ONLINE       rac2                     STABLE
ora.ons
                ONLINE  ONLINE       rac1                     STABLE
                ONLINE  ONLINE       rac2                     STABLE
--------------------------------------------------------------------------------
Cluster Resources
--------------------------------------------------------------------------------
ora.LISTENER_SCAN1.lsnr
      1        ONLINE  ONLINE       rac1                     STABLE
ora.MGMTLSNR
      1        ONLINE  ONLINE       rac1
169.254.73.59 192.16
                                                          8.2.101,STABLE
ora.asm
      1        ONLINE  ONLINE       rac1
Started,STABLE
      2        ONLINE  ONLINE       rac2
Started,STABLE
      3        OFFLINE OFFLINE                               STABLE
ora.cndba.cndbaapp.svc
      1        ONLINE  ONLINE       rac1                     STABLE
ora.cndba.cndbaapp2.svc
      1        ONLINE  ONLINE       rac1                     STABLE
      2        ONLINE  ONLINE       rac2                     STABLE
ora.cndba.db
      1        ONLINE  ONLINE       rac1
Open,HOME=/u01/app/o

racle/product/18.1.0
                                                          /db_1,STABLE
      2        ONLINE  ONLINE       rac2
Open,HOME=/u01/app/o

racle/product/18.1.0
                                                          /db_1,STABLE
ora.cvu
      1        ONLINE  ONLINE       rac1                     STABLE
ora.gns
      1        ONLINE  ONLINE       rac1                     STABLE
ora.gns.vip
      1        ONLINE  ONLINE       rac1                     STABLE
ora.mgmtdb
```

```
     1        ONLINE  ONLINE       rac1                      Open,STABLE
ora.qosmserver
     1        ONLINE  ONLINE       rac1                      STABLE
ora.rac1.vip
     1        ONLINE  ONLINE       rac1                      STABLE
ora.rac2.vip
     1        ONLINE  ONLINE       rac2                      STABLE
ora.scan1.vip
     1        ONLINE  ONLINE       rac1                      STABLE
-------------------------------------------------------------------------
```

（12）验证整个集群各个节点 OCR 的完整性，用 grid 账户运行以下命令：

```
[grid@rac1 ~]$ cluvfy comp ocr -n all -verbose
Verifying OCR Integrity ...PASSED
Verification of OCR integrity was successful.
CVU operation performed:     OCR integrity
Date:                       Jul 27, 2018 10:13:12 AM
CVU home:                   /u01/app/18.1.0/grid/
User:                       grid
```

13.5.5 使用导出和导入恢复 OCR

除可以使用 OCR 的自动备份外，还可以使用导出和导入的方式进行备份。注意，使用 ocrconfig -export 命令导出的 OCR 备份只能使用 ocrconfig-import 命令来导入，不能使用 ocrconfig-restore 命令来恢复。

另外，ocrconfig backup 命令是 OCR 的一致快照，是在 system online 时创建的，而 export 不是一致的快照。如果想使用 ocrconfig -export 命令来执行一致的快照，就需要关闭所有节点的集群。使用导出和导入的方式恢复 OCR 是不推荐的，读者在这里了解一下即可。

以下命令如无特殊说明，都用 root 账户执行。

备份 OCR，代码如下：

```
    [root@rac1 ~]# ocrconfig -export /u01/ocr_20180913
    PROT-58: successfully exported the Oracle Cluster Registry
contents to file '/u01/ocr_20180913'
    [root@rac1 ~]#
```

恢复 OCR 时先关闭 CRS，并以独占模式启动，代码如下：

```
./ crsctl stop crs
./ crsctl start crs -excl
```

导入 OCR 备份，代码如下：

```
./ ocrconfig -import /u01/ocr_20180913
```

验证 OCR 的完整性，重启时以正常模式启动 CRS，代码如下：

```
./ocrcheck
./crsctl stop crs
./crsctl start crs
```

用 grid 账户来验证集群中所有节点的 OCR 完整性，代码如下：

```
cluvfy comp ocr -n all -verbose
```

13.5.6　Oracle 本地注册表（OLR）

在 RAC 集群中的每个节点上都有一个记录本节点特定资源的 OLR（Oracle Local Registry），在安装 OCR 的同时也安装和配置 OLR。本地节点上的进程都可以对 OLR 进行同步读写。OLR 默认位于$GIRD_HOME/cdata/host_name.olr。可以通过 OCRCHECK、OCRDUMP 和 OCRCONFIG 命令加上参数-local 来管理 OLR。

1．查看本节点 OLR 的状态

查看本节点 OLR 的状态的代码如下：

```
[root@rac1 bin]# ./ocrcheck -local
Status of Oracle Local Registry is as follows :
       Version                  :  4
       Total space (kbytes)     :    409568
       Used space (kbytes)      :      1096
       Available space (kbytes) :    408472
       ID                     :  117651951
       Device/File Name         :
/u01/app/18.1.0/grid/cdata/rac1.olr
                                Device/File integrity check
succeeded
       Local registry integrity check succeeded
       Logical corruption check succeeded
```

2．查看 OLR 中的内容

使用 OCRDUMP 命令可以将 OLR 中的内容输出，代码如下：

```
[root@rac1 bin]# ./ocrdump -local -stdout
07/27/2018 11:00:16
./ocrdump.bin -local -stdout

[SYSTEM]
UNDEF :
SECURITY : {USER_PERMISSION : PROCR_ALL_ACCESS, GROUP_PERMISSION :
PROCR_READ, OTHER_PERMISSION : PROCR_READ, USER_NAME : root,
GROUP_NAME : root}
```

3. 管理 OLR

使用 OCRCONFIG 命令可以管理 OLR，包括备份、恢复等。

- 导出 OLR，代码如下：

```
[root@rac1 bin]# ./ocrconfig -local -export
/u01/app/18.1.0/grid/cdata/olr_2018_07_27_export.bak
PROTL-58: successfully exported the Oracle Local Registry contents
to file '/u01/app/18.1.0/grid/cdata/olr_2018_07_27_export.bak'
[root@rac1 bin]# ll
/u01/app/18.1.0/grid/cdata/olr_2018_07_27_export.bak
-rw------- 1 root root 69632 Jul 27 11:03
/u01/app/18.1.0/grid/cdata/olr_2018_07_27_export.bak
```

- 导入 OLR，代码如下：

```
[root@rac1 bin]# ./ocrconfig -local -import
/u01/app/18.1.0/grid/cdata/olr_2018_07_27_export.bak
```

要关闭 CRS 才能进行导入操作，否则会报以下错误。

```
PROTL-19: Cannot proceed while the Oracle High Availability Service
is running
```

- 手动备份 OLR。和备份 OCR 一样，除可以通过导出来备份外，还可以手动备份。注意，Oracle 不会自动备份 OLR。

```
[root@rac1 bin]# ./ocrconfig -local -manualbackup
rac1     2018/07/27 11:06:52
/u01/app/18.1.0/grid/cdata/rac1/backup_20180727_110652.olr     0
rac1     2018/07/24 15:00:00
/u01/app/18.1.0/grid/cdata/rac1/backup_20180724_150000.olr     0
```

- 查看 OLR 备份，代码如下：

```
[root@rac1 bin]# ./ocrconfig -local -showbackup
```

```
    rac1     2018/07/27 11:06:52
/u01/app/18.1.0/grid/cdata/rac1/backup_20180727_110652.olr    0
    rac1     2018/07/24 15:00:00
/u01/app/18.1.0/grid/cdata/rac1/backup_20180724_150000.olr    0
```

- 查看 OLR 备份文件的内容，代码如下：

```
    [root@rac1 bin]# ./ocrdump -local -backupfile
/u01/app/18.1.0/grid/cdata/rac1/backup_20180727_110652.olr
```

- 修改 OLR 备份路径，代码如下：

```
    [root@rac1 bin]# ./ocrconfig -local -backuploc /data/olrbackup/
```

- 恢复 OLR。先关闭 CRS，然后通过备份恢复 OLR，代码如下：

```
    [root@rac1 bin]# ./crsctl stop crs
    [root@rac1 bin]# ./ocrconfig -local -restore
/u01/app/18.1.0/grid/cdata/rac1/backup_20180727_110652.olr
    [root@rac1 bin]# ./ocrcheck -local
    [root@rac1 bin]# ./crsctl start crs
    [grid@rac1 bin]$ ./ cluvfy comp olr
```

13.6　管理 Voting File

Voting File 中存储了 RAC 节点的成员信息，每个节点在启动并试图加入 RAC 集群时，都需要读取 Voting File，以确定当前节点的成员资格。Voting Disk 是为了在出现脑裂时，决定哪个节点获得控制权，其他的节点必须从集群中被剔除。Voting File 对集群来说非常重要，所以在安装集群时需要创建多个 Voting File，每个 Voting File 都有唯一的 ID 即 FUID（File Universal Identifier）。

Voting File 存放在 OCR 的磁盘组中，当 OCR 磁盘组的冗余类型确定后，用户就无法手动修改 Voting File 的数量，统一由 Oracle 自动管理。

13.6.1　备份 Voting File

在 Oracle 12c 之前，可以使用 dd 命令对 Voting Disk 进行备份和恢复，但从 Oracle 12c 开始已经不支持 dd 命令了。现在 Oracle 在备份 OCR 时会自动备份 Voting File，所以无须单独备份 Voting File。

13.6.2 恢复 Voting File

当 Voting File 所在的磁盘组中的故障组的损坏数量超过了冗余所允许的范围时，CRS 是无法正常启动的，必须进行恢复才能使 CRS 正常启动，但是在恢复 Voting File 之前，要确保 OCR 是正常的，如果 OCR 也出现了故障，那么就需要先恢复 OCR，再恢复 Voting File。

恢复 Voting File 的操作步骤如下。

（1）查看 Voting File 的信息，代码如下：

```
[root@rac1 bin]# ./crsctl query css votedisk
## STATE    File Universal Id                File Name Disk group
-- -----    -----------------                --------- ---------
 1. ONLINE   f2596b60ae784f3fbf8eec615b504c94 (/dev/asmdiskg)
[OCR_MGMT]
 2. ONLINE   b67888fbed364feebf529b6085c27fa2 (/dev/asmdiskf)
[OCR_MGMT]
 3. ONLINE   b2dd4cdaa3804f13bf35a41d3bb55da2 (/dev/asmdiske)
[OCR_MGMT]
Located 3 voting disk (s) .
```

（2）关闭 CRS（所有节点），代码如下：

```
[root@rac1 bin]#./crsctl stop crs [-f]
```

如果无法正常关闭 CRS，则加参数-f 强制关闭 CRS。

（3）以独占模式打开 CRS，代码如下：

```
crsctl start crs -excl
```

（4）恢复 Voting File，代码如下：

```
[root@rac1 bin]# ./crsctl replace votedisk +ocr_test
Successful addition of voting disk
03b14cc9a8784f59bf500471a2025d83.
Successful deletion of voting disk
f2596b60ae784f3fbf8eec615b504c94.
Successful deletion of voting disk
b67888fbed364feebf529b6085c27fa2.
Successful deletion of voting disk
b2dd4cdaa3804f13bf35a41d3bb55da2.
Successfully replaced voting disk group with +ocr_test.
CRS-4266: Voting file (s) successfully replaced
```

注意这里在恢复时用的是 replace 命令，该命令除用于恢复外，还可以用于迁移。

（5）再次查看 Voting File，代码如下：

```
[root@rac1 bin]# ./crsctl query css votedisk
##  STATE    File Universal Id                File Name Disk group
--  -----    -----------------                --------- ---------
 1. ONLINE   03b14cc9a8784f59bf500471a2025d83 (/dev/asmdiskh)
[OCR_TEST]
Located 1 voting disk(s).
```

（6）关闭 CRS 并正常启动，代码如下：

```
[root@rac1 bin]# ./crsctl stop crs
[root@rac1 bin]# ./crsctl start crs
```

第 14 章

RAC 的负载均衡

负载均衡（Load Balance）是 Oracle RAC 最重要的特性之一，主要是把负载平均分配到集群中的各个节点，以提高系统的整体吞吐能力。通常情况下有两种方式来实现负载均衡，一个是客户端的负载均衡，另一个是服务端的负载均衡。

14.1 客户端均衡

客户端均衡（Client-Side Load Balance）的配置方法是在客户端的 tnsnames.ora 文件中加入 LOAD_BALANCE=ON。在没有使用 SCAN IP 时，当客户端发起连接后，会从地址列表中随机选取一个 IP 地址，再使用随机算法把连接请求分配到各个实例。算法如下：

```
RAC =
  (DESCRIPTION =
    (ADDRESS = (PROTOCOL = TCP)(HOST = rac1-vip)(PORT = 1521))
    (ADDRESS = (PROTOCOL = TCP)(HOST = rac2-vip)(PORT = 1521))
    (LOAD_BALANCE = ON)
    (CONNECT_DATA =
      (SERVER = DEDICATED)
      (SERVICE_NAME = RAC)
    )
   )
  )
```

tnsnames.ora 文件比较敏感，手敲代码容易出错，建议使用 Netmgr 工具进行配置。另外，这种随机算法没有考虑每个节点的真实负载，所以分配结果不一定是均衡的，还需要依赖服务端均衡（Sevice-Side Load Balance）。

14.2　服务端均衡（通过监听器）

服务端均衡依赖监听器（Listener）收集的负载信息。在数据库运行过程中，PMON 后台进程会收集系统的负载信息，然后登记到监听器中。最少 1 分钟、最多 10 分钟 PMON 就要做一次信息更新，并且节点的负载越高，更新频率就越高，以保证监听器能掌握每个节点准确的负载情况。如果监听器被关闭了，则 PMON 进程会每隔 1 秒检查一次监听器是否重启。除这个自动的、定时的更新任务外，用户也可以使用 alter system register 命令手动执行这个过程。

可以从监听日志中查看注册信息，实例启动时 PMON 进程进行的第一次登记过程叫作 service-register，之后的更新过程叫作 service-update，代码如下：

```
[oracle@rac1 trace]$ cat listener.log |grep service_register
05-SEP-2018 00:47:31 * (ADDRESS=(PROTOCOL=ipc)(KEY=LISTENER))
* service_register * LsnrAgt * 0
05-SEP-2018 00:48:29 * (ADDRESS=(PROTOCOL=tcp)
(HOST=192.168.1.100)(PORT=57909)) * service_register * +ASM1 * 0
05-SEP-2018 00:52:57 * (ADDRESS=(PROTOCOL=tcp)
(HOST=192.168.1.100)(PORT=35804)) * service_register * -MGMTDB * 0
05-SEP-2018 00:54:01 * (ADDRESS=(PROTOCOL=tcp)
(HOST=192.168.1.100)(PORT=35820)) * service_register * -MGMTDB * 0
05-SEP-2018 00:54:36 * (ADDRESS=(PROTOCOL=tcp)
(HOST=192.168.1.100)(PORT=35848)) * service_register * -MGMTDB * 0
05-SEP-2018 00:56:47 * (ADDRESS=(PROTOCOL=tcp)
(HOST=192.168.1.100)(PORT=35882)) * service_register * -MGMTDB * 0
05-SEP-2018 01:03:26 * (ADDRESS=(PROTOCOL=tcp)
(HOST=192.168.1.100)(PORT=35966)) * service_register * -MGMTDB * 0
05-SEP-2018 01:09:59 * (ADDRESS=(PROTOCOL=tcp)
(HOST=192.168.1.100)(PORT=36126)) * service_register * -MGMTDB * 0
…

[oracle@rac1 trace]$ tail -200 listener.log |grep service_update
14-SEP-2018 14:17:59 * service_update * +ASM1 * 0
14-SEP-2018 14:18:35 * service_update * +ASM1 * 0
14-SEP-2018 14:19:05 * service_update * cndba1 * 0
14-SEP-2018 14:19:14 * service_update * cndba1 * 0
14-SEP-2018 14:22:11 * service_update * cndba1 * 0
14-SEP-2018 14:22:14 * service_update * cndba1 * 0
…
```

PMON 进程不仅会在本地的监听器中注册，还会在其他节点的监听器中注册。对于服务端负载均衡，可以手动配置remote_listener，Oracle建议将其设置为scan_name:scan_port，代码如下：

```
SQL> show parameter _listener

NAME                    TYPE        VALUE
---------------------------------------------------------------
forward_listener            string
local_listener                   string           (ADDRESS=
(PROTOCOL=TCP)(HOST=192.168.1.103)(PORT=1521))
remote_listener             string          rac-scan:1521
```

可以看到，这里的remote_listener 用的是SCAN（Single Client Access Name）的地址。从Oracle 11g 开始，在安装RAC 时都需要使用SCAN，而SCAN的配置又依赖DNS，在没有DNS的情况下，一般采用另一种方式，即将SCAN IP写入/etc/hosts 文件，这种方式可以满足RAC 的安装要求，但达不到SCAN 一对多个IP的要求。在这种情况下，就需要在tnsnames.ora 文件中单独配置2个VIP作为remote_listener。

有了PMON的自动注册机制后，集群的每个节点的监听器都掌握所有节点的负载情况，当收到客户端连接请求时，就会把连接转给负载最小的节点，这个节点有可能是自己，也有可能是其他节点，也就是监听器会转发用户的请求。

Oracle 11g 以后的版本监听都是在GI用户下的，这里修改Oracle用户的tnsnames.ora 文件，在所有节点的该文件中添加如下内容：

```
 [oracle@rac1 admin]$ cat tnsnames.ora
rac1_local=
  (ADDRESS = (PROTOCOL = TCP)(HOST = rac1-vip)(PORT = 1521))

rac2_local=
  (ADDRESS = (PROTOCOL = TCP)(HOST = rac2-vip)(PORT = 1521))

rac_remote =
  (DESCRIPTION =
    (ADDRESS = (PROTOCOL = TCP)(HOST = rac1-vip)(PORT = 1521))
    (ADDRESS = (PROTOCOL = TCP)(HOST = rac2-vip)(PORT = 1521))
  )
```

然后修改参数，代码如下：

```
SQL> alter system set remote_listener='rac_remote' scope=both
```

```
sid='*';
    System altered.
```

可以看出，如果使用 DNS 配置了 SCAN IP，那么就不再需要配置了，默认就可以满足；仅当 SCAN IP 被写入/etc/hosts 的情况下，才需要做以上配置。

14.3　服务端均衡（通过服务）

14.2 节通过配置 remote_listener 实现了服务端的负载均衡，但 Oracle 不推荐使用这种方式，Oracle 建议通过服务（Service）来实现。如果非要使用 remote_listener，建议将其设置为 scan_name:scan_port。

通过服务可以实现自动的负载均衡，而且更灵活。通过服务，可以为不同的应用程序添加不同的服务，而不同的服务具有不同的配置，如首选节点、备用节点、重连次数、重连时间间隔等。

服务中的负载均衡有以下两种类型。

- SHORT：用于运行时负载均衡的应用程序，通常指通过连接池来连接数据库的应用，如 OLTP 系统。
- LONG：默认值，用于典型的批处理操作，如 OLAP 系统。

在创建或修改服务时添加参数-clbgoal 来指定服务的类型就可以了，代码如下：

```
    $ srvctl modify service -db db_unique_name -service oltpapp
-clbgoal SHORT
    $ srvctl modify service -db db_unique_name -service batchconn
-clbgoal LONG
```

实际上，这里就是利用服务将会话连接到不同的节点，是故障转移配置中的一个选项，该部分的测试可参考第 15 章故障转移部分，这里只需要记住对于 OLTP 使用 SHORT 类型，而对于 OLAP 使用 LONG 类型。

第 15 章

RAC 的故障转移

故障转移（Failover）指集群中任何一个节点发生故障都不会影响用户的使用，连接到故障节点的会话会被自动转移到正常的节点。

Oracle RAC 的故障转移可以分为以下 3 种。

- 客户端连接时故障转移。
- 客户端 TAF。
- 服务端 TAF

15.1 客户端连接时故障转移

如果用户端 tnsnames.ora 文件中配置了多个地址，则用户在发起连接请求时，会先尝试连接地址表中的第一个地址，如果这个连接失败，则继续尝试连接第二个地址，直至连接成功或遍历完所有的地址。

这种故障转移方式只有在发起连接时才会去感知节点故障，如果节点没有反应，则自动尝试连接地址列表中的下一个地址。连接建立之后，即使节点出现故障也不会进行处理，会话中断，应用必须重新建立连接。这种故障转移方式在 tnsnames.ora 文件中添加 FAILOVER=ON 即可实现，默认是启用的，不添加该条目也可以使用该功能。

通过 SCAN IP 的示例代码如下：

```
CNDBA =
  (DESCRIPTION =
    (ADDRESS = (PROTOCOL = TCP)(HOST = rac-scan)(PORT = 1521))
    (CONNECT_DATA =
      (SERVER = DEDICATED)
      (SERVICE_NAME = cndba)
    )
```

```
)
```

通过 VIP 的示例代码如下：

```
CNDBA =
  (DESCRIPTION =
    (ADDRESS = (PROTOCOL = TCP)(HOST = rac1-vip)(PORT = 1521))
    (ADDRESS = (PROTOCOL = TCP)(HOST = rac2-vip)(PORT = 1521))
    (CONNECT_DATA =
      (SERVER = DEDICATED)
      (SERVICE_NAME = cndba)
    )
  )
```

15.2　客户端 TAF

TAF（Transparent Application Failover）指建立连接后，如果某个实例发生故障，则连接到该实例上的会话会自动迁移到其他正常的实例上。对于应用程序而言，这个迁移过程是透明的，不需要用户的介入。TAF 的配置也很简单，只需要在客户端的 tnsnames.ora 文件中添加 FAILOVER_MODE 配置项，再配置以下 4 个参数。

- METHOD：用户定义何时创建到其他实例的连接，有 BASIC 和 PRECONNECT 两个可选值。
  - BASIC：在感知到节点故障时，创建到其他实例的连接。
  - PRECONNECT：在最初建立连接时就建立到所有实例的连接，当发生故障时，立刻就可以切换到其他链路上。

  BASIC 方式在故障转移时会有时间延迟，但节省资源，并且 RAC 出现故障的概率较小，所以一般选择 BASIC 方式。
- TYPE：用于定义发生故障时对完成的 SQL 语句的处理，有三种处理方式，即 session、select 和 none（默认值）。前面两种方式对未提交的事务都会自动回滚，区别在于对 select 语句的处理。对于 select 方式，用户正在执行的 select 语句会被转移到新的实例上，在新的节点上继续返回后续结果集，而已经返回的记录集则被抛弃。为了实现 select 方式，Oracle 必须为每个会话保存更多的内容，包括游标、用户上下文等，用资源换时间。
- DELAY：重新连接的时间间隔。

- RETRIES：重新连接的次数。

具体示例代码如下：

```
CNDBA_TAF =
  (DESCRIPTION =
    (ADDRESS = (PROTOCOL = TCP)(HOST = rac1-vip)(PORT = 1521))
    (ADDRESS = (PROTOCOL = TCP)(HOST = rac2-vip)(PORT = 1521))
    (
 CONNECT_DATA=
    (SERVER=DEDICATED)
    (SERVICE_NAME=CNDBA)
    (
    FAILOVER_MODE=
      (TYPE=session)
      (METHOD=basic)
      (RETRIES=180)
      (DELAY=5)
    )
    )
  )
```

15.3 服务端 TAF

当应用程序客户端较多时，配置客户端 TAF 就会比较麻烦，因为客户端 TAF 是通过 tnsnames.ora 文件来配置的。服务端 TAF 是在数据库服务器上进行配置的，把所有的 TAF 的 FAIL_MODE 配置保存在数据字典中，就省去了客户端的配置工作。

服务端 TAF 比客户端 TAF 多了一个 Instance Role（实例角色）的概念。当有多个实例时，可以配置优先使用哪一个实例提供服务，有以下两种类型。

- PREFERRED：首选实例，会优先选择拥有这个角色的实例提供服务。
- AVAILABLE：后备实例。

应用程序优先连接到 PREFFERRED 配置的实例上，当 PREFERRED 配置的实例不可用时，才会被转到 AVAILBALE 配置的备用实例上。

15.4　服务端 TAF 配置示例

下面通过具体的例子演示如何配置服务来实现服务端的 TAF。演示的环境有两个节点，即 rac1 和 rac2，其实例名分别是 cndba1 和 cndba2。

（1）通过 SRVCTL 命令添加一个服务，用于 TAF。注意，要使用 Oracle 账户来添加该服务。SRVCTL 命令的选项很多，可以查看帮助文档了解详细用法，输出内容也较多，这里只列出命令：

```
[oracle@rac1 ~]$ srvctl add service -h
```

使用命令帮助功能会输出所有的参数，代码如下：

```
    srvctl add service -d <db_unique_name> -s <service_name>
    -r "<preferred_list>" [-a "<available_list>"] [-P {BASIC | NONE
| PRECONNECT}]
    -g <server_pool> [-c {UNIFORM | SINGLETON}]
    [-k <net_num>]
    [-l
[PRIMARY][,PHYSICAL_STANDBY][,LOGICAL_STANDBY][,SNAPSHOT_STANDBY]]
    [-y {AUTOMATIC | MANUAL}]
    [-q {TRUE|FALSE}]
    [-x {TRUE|FALSE}]
    [-j {SHORT|LONG}]
    [-B {NONE|SERVICE_TIME|THROUGHPUT}]
    [-e {NONE|SESSION|SELECT}]
    [-m {NONE|BASIC}]
    [-z <failover_retries>]
    [-w <failover_delay>]

    [oracle@rac1 ~]$ srvctl add service -d cndba -s cndba_taf -r
"cndba1" -a "cndba2" -j short -P basic -e select -m basic -w 5 -z 180
-q true
```

（2）启用服务，代码如下：

```
[oracle@rac1 ~]$ srvctl start service -d cndba -s cndba_taf
```

（3）查看服务状态，代码如下：

```
[oracle@rac1 ~]$ srvctl config service -d cndba
Service name: cndba_taf
Server pool:
Cardinality: 1
```

```
Service role: PRIMARY
Management policy: AUTOMATIC
DTP transaction: false
AQ HA notifications: true
Global: false
Commit Outcome: false
Failover type: SELECT
Failover method: BASIC
Failover retries: 180
Failover delay: 5
Failover restore: NONE
Connection Load Balancing Goal: SHORT
Runtime Load Balancing Goal: NONE
TAF policy specification: BASIC
Edition:
Pluggable database name:
Hub service:
Maximum lag time: ANY
SQL Translation Profile:
Retention: 86400 seconds
Replay Initiation Time: 300 seconds
Drain timeout:
Stop option:
Session State Consistency: DYNAMIC
GSM Flags: 0
Service is enabled
Preferred instances: cndba1
Available instances: cndba2
CSS critical: no
[oracle@rac1 ~]$
```

注意，这里配置了 Connection Load Balancing Goal: SHORT，对应的是 OLTP 系统。

（4）可以通过 dba_services 视图获取相关信息，代码如下：

```
SQL> set lines 120
SQL> col name for a15
SQL> select name,service_id from dba_services where name
='cndba_taf';

NAME    SERVICE_ID
--------------- ----------
cndba_taf        3
```

```
SQL>col name format a15
SQL>col failover_method format a11 heading 'METHOD'
SQL>col failover_type format a10 heading 'TYPE'
SQL>col failover_retries format 9999999 heading 'RETRIES'
SQL>col goal format a10
SQL>col clb_goal format a8
SQL>col AQ_HA_NOTIFICATIONS format a5 heading 'AQNOT'

SQL>select name, failover_method, failover_type,
failover_retries,goal, clb_goal,aq_ha_notifications
from dba_services where service_id = 3;
NAME      METHOD      TYPE RETRIES GOAL   CLB_GOAL AQNOT
-------------------- ---------- -------- ---------- -------- -----
cndba_taf    BASIC        SELECT        180 NONE     SHORT    YES
```

（5）查看监听服务是否被注册，代码如下：

```
[oracle@rac1 ~]$ lsnrctl status
…
Service "cndba_taf" has 1 instance(s).
  Instance "cndba1", status READY, has 2 handler(s) for this
service...
…
The command completed successfully
```

（6）测试 TAF ，在 tnsnames.ora 文件中进行配置即可，代码如下：

```
[oracle@rac2 admin]$ cat tnsnames.ora
cndba_taf =
  (DESCRIPTION =
    (ADDRESS = (PROTOCOL = TCP)(HOST = rac-scan)(PORT = 1521))
    (CONNECT_DATA =
      (SERVER = DEDICATED)
      (SERVICE_NAME = cndba_taf)
    )
  )
```

检查 Service 参数，如果没有刚才创建的 Service，则手工添加，否则使用该 Service 时会报如下错误：

```
ORA-12514: TNS:listener does not currently know of service
requested in connect descriptor
SQL> show parameter service
```

```
NAME                    TYPE          VALUE
------------------------------- ------------------------------
service_names                     string          cndba
```

这里 Service 参数没有更新，手工更新一下，代码如下：

```
    SQL> alter system set service_names='cndba_taf', 'cndba'
scope=both sid='*';
    System altered.

    SQL> show parameter service
    NAME                    TYPE          VALUE
    ------------------------------- ------------------------------
    service_names                     string          cndba_taf, cndba

    [oracle@rac2 admin]$ sqlplus sys/oracle@cndba_taf as sysdba

    SQL*Plus: Release 18.0.0.0.0 - Production on Fri Sep 14 20:01:00
2018
    Version 18.3.0.0.0

    Copyright (c) 1982, 2018, Oracle.  All rights reserved.


    Connected to:
    Oracle Database 18c Enterprise Edition Release 18.0.0.0.0 -
Production
    Version 18.3.0.0.0

    SQL> set lines 120
    SQL> col host_name for a15
    SQL> col instance_name for a15
    SQL> select host_name,instance_name from v$instance;

    HOST_NAME    INSTANCE_NAME
    --------------- ---------------
    rac1    cndba1
```

关闭实例 2，注意不能以正常方式去关闭，必须以 abort 方式关闭，代码如下：

```
    [grid@rac1 admin]$ srvctl stop instance -d cndba -i cndba1
-stopoption abort -force
    [grid@rac1 admin]$
```

再次查看，当前会话没有被中断，并且成功地转移到节点 2 了，代码如下：

```
SQL> select host_name,instance_name from v$instance;

HOST_NAME     INSTANCE_NAME
--------------- ---------------
rac2     cndba2
```

（7）之前的配置是直接连接 CDB 的，代码如下：

```
SQL> show pdbs

   CON_ID CON_NAME              OPEN MODE  RESTRICTED
---------- ------------------------------ ---------- ----------
     2 PDB$SEED              READ ONLY  NO
     3 DAVE            MOUNTED
SQL> show con_name

CON_NAME
------------------------------
CDB$ROOT
```

可以在 Service 参数中加-pdb 来连接 PDB 库，代码如下：

```
[oracle@rac1 ~]$ srvctl add service -d cndba -s pdb_taf -r "cndba1"
-a "cndba2" -pdb "dave" -j short -P basic -e select -m basic -w 5 -z
180 -q true

[oracle@rac1 ~]$ srvctl config service -d cndba -s pdb_taf
Service name: pdb_taf
…
Edition:
Pluggable database name: dave
Preferred instances: cndba1
Available instances: cndba2
…

[oracle@rac2 ~]$ sqlplus sys/oracle@pdb_taf as sysdba

SQL*Plus: Release 18.0.0.0.0 - Production on Fri Sep 14 20:30:26
2018
Version 18.3.0.0.0
Copyright (c) 1982, 2018, Oracle.  All rights reserved.
```

```
    Connected to:
    Oracle Database 18c Enterprise Edition Release 18.0.0.0.0 -
Production
    Version 18.3.0.0.0

    SQL> show con_name

    CON_NAME
    ------------------------------
    DAVE
    SQL> show pdbs

    CON_ID CON_NAME            OPEN MODE  RESTRICTED
    ---------- ------------------------------ ---------- ----------
    3 DAVE             READ WRITE NO
    SQL> select name,pdb from v$services order by 1;

    NAME         PDB
    -------------------- --------------------
    SYS$BACKGROUND      CDB$ROOT
    SYS$USERS         CDB$ROOT
    cndba             CDB$ROOT
    cndbaXDB     CDB$ROOT
    cndba_taf         CDB$ROOT
    dave         DAVE
    pdb_taf      CDB$ROOT
```

在测试之前要确保 Service 是启动的，如果没有启动，则会直接连接到 CDB，而不是 PDB。另外，要保证 service_names 参数中要包含参数，否则报 ORA-12514 错误。

第 16 章 RAC 中的 GIMR

16.1 GIMR 概述

Oracle 12.1.0.1 中引入了 GIMR（Grid Infrastructure Management Repository），在安装 GI 时会提示用户是否配置 GIMR，如果选择安装，则会创建一个 MGMT 数据库。但从 Oracle 12.1.0.2 开始，在安装 GI 时已经默认安装 GIMR 了，如图 16-1 所示。

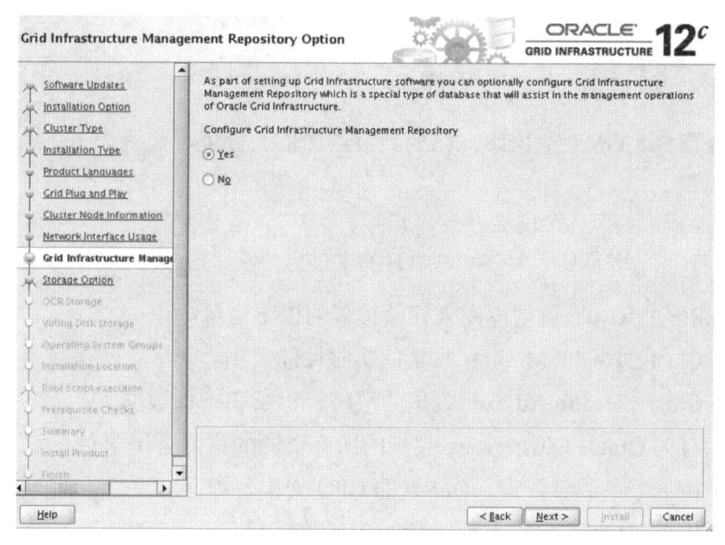

图 16-1　GI 的安装界面

GIMR 必须运行在 Hub 节点上，Oracle 推荐将其数据文件存储在单独的磁盘组中（MGMT），如果不指定单独的磁盘组，那么默认和 OCR 存储在同一个磁盘组中。GIMR 通过专用网络与任何集群客户端（如 Cluster Health Advisor、Rapid Home Provisioning Server、OLOGGERD 和 OCLUMON）进行通信；通过公共网络，GIMR 与外部客户端、域服务集群和集群成员进行通信。

MGMT 数据库是 GIMR 的存储库，是受 RAC 集群管理的一个单实例，在集群启动时会在其中一个节点上运行，并接受 GI 的管理。如果运行 MGMT 数据库的节点宕机了，则 GI 会自动把 MGMT 数据库转移到其他的节点上。默认情况下，MGMT 数据库中的数据文件被存放在共享的设备上（如 OCR/Voting Files 的磁盘组中），可以移动其位置。

查看当前 MGMT 数据库运行的节点，代码如下：

```
[grid@rac1 ~]$ srvctl status mgmtdb
Database is enabled
Instance -MGMTDB is running on node rac1
```

关闭该节点的 CRS，代码如下：

```
[root@RAC1 ~]# cd /u01/app/18.3.0/grid/bin/
[root@RAC1 bin]# ./crsctl stop crs
CRS-2672: Attempting to start 'ora.mgmtdb' on 'rac2'
CRS-2676: Start of 'ora.qosmserver' on 'rac2' succeeded
CRS-2676: Start of 'ora.mgmtdb' on 'rac2' succeeded
```

再次查看 MGMT 数据库，已经到节点 2 上运行了，代码如下：

```
[grid@rac2 ~]$ srvctl status mgmtdb
Database is enabled
Instance -MGMTDB is running on node rac2
```

GIMR 的 MGMT 数据库用来存储集群的以下信息。

- Cluster Health Monitor 收集的实时性能数据。
- Cluster Health Advisor 收集的故障、诊断和指标数据。
- 有关 Oracle Clusterware 收集的所有资源的集群范围事件。
- 用于服务质量管理（QoS）的 CPU 架构数据。
- Rapid Home Provisioning 所需的元数据。

16.2 MGMT 数据库

MGMT 数据库是一个单实例，实际上，MGMT 数据库是带一个 PDB 的

CDB 库，我们可以使用 GI 中的命令直接操作 MGMT 数据库对应的 PDB。

查看 MGMT 数据库当前的节点，代码如下：

```
[grid@rac2 ~]$ oclumon manage -get master
Master = rac1
[grid@rac2 ~]$
```

查看状态，代码如下：

```
Master = rac1
[grid@rac2 ~]$ srvctl status mgmtdb
Database is enabled
Instance -MGMTDB is running on node rac1
```

查看配置信息，代码如下：

```
[grid@rac1 ~]$ srvctl config mgmtdb
Database unique name: _mgmtdb
Database name:
Oracle home: <CRS home>
Oracle user: grid
Spfile: +MGMT/_MGMTDB/PARAMETERFILE/spfile.270.986000873
Password file:
Domain:
Start options: open
Stop options: immediate
Database role: PRIMARY
Management policy: AUTOMATIC
Type: Management
PDB name: GIMR_DSCREP_10
PDB service: GIMR_DSCREP_10
Cluster name: rac
Database instance: -MGMTDB
```

连接 MGMT 数据库实例，代码如下：

```
[grid@rac1 ~]$ export ORACLE_SID=-MGMTDB
[grid@rac1 ~]$ sqlplus / as sysdba
SQL*Plus: Release 18.0.0.0.0 - Production on Thu Sep 13 15:37:05
2018
Version 18.3.0.0.0
Copyright (c) 1982, 2018, Oracle. All rights reserved.
Connected to:
Oracle Database 18c Enterprise Edition Release 18.0.0.0.0 -
Production
```

```
Version 18.3.0.0.0
SQL> show pdbs;
   CON_ID CON_NAME                       OPEN MODE  RESTRICTED
---------- ------------------------------ ---------- ----------
         2 PDB$SEED                      READ ONLY  NO
         3 GIMR_DSCREP_10                 READ WRITE NO
SQL> select name,cdb from v$database;
NAME  CDB
--------- ---
_MGMTDB   YES
SQL> select file_name from dba_data_files union select member
file_name from V$logfile;
FILE_NAME
--------------------------------------------------------------------
------------------
+MGMT/_MGMTDB/DATAFILE/sysaux.259.986000013
+MGMT/_MGMTDB/DATAFILE/system.258.985999993
+MGMT/_MGMTDB/DATAFILE/undotbs1.260.986000029
+MGMT/_MGMTDB/ONLINELOG/group_1.264.986000083
+MGMT/_MGMTDB/ONLINELOG/group_2.262.986000083
+MGMT/_MGMTDB/ONLINELOG/group_3.263.986000083
6 rows selected.
```

可以看到这里用的磁盘组是+MGMT，这是单独创建的磁盘组，可以通过asmcmd命令进行查看，代码如下：

```
[grid@rac1 trace]$ asmcmd lsdg
    State      Type      Rebal   Sector   Logical_Sector   Block          AU
Total_MB    Free_MB    Req_mir_free_MB    Usable_file_MB    Offline_disks
Voting_files   Name
MOUNTED  NORMAL  N        512           512   4096  4194304
61440    41320            0         20660          0            N
DATA/
MOUNTED  EXTERN  N        512           512   4096  4194304
40960    35684            0         35684          0            N
MGMT/
MOUNTED  NORMAL  N        512           512   4096  4194304
6144     5228          2048          1590          0            Y
OCR/
[grid@rac1 trace]$
```

通过查看官方手册，可知在Oracle 18c中，如果磁盘组中只需要存放OCR

和 Voting Files，则对磁盘组的磁盘大小最低要求是 2GB，但因为 GIMR 默认也是存放在OCR磁盘组中的，所以就要求OCR 磁盘组的磁盘大小最小为30GB。为了方便管理，建议将 OCR 和 GIMR 存放在不同的磁盘组中。

16.3　MGMT 数据库的管理

管理 MGMT 数据库只需要两个命令：SRVCTL...MGMTDB 和 SERVCTL...MGMTLSNR。理论上对 MGMT 数据库是不需要手动管理的（包括优化、备份等），都是由 Oracle 自动管理。但是如果有必要，可以进行手动管理。

16.3.1　关闭、启动 MGMT 数据库

MGMT 数据库会随着集群的启动而自动启动，也可以通过 SRVCTL START MGMTDB 命令和 SRVCTL STOP MGMTDB 命令来启动和关闭数据库，语法如下：

```
    srvctl start mgmtdb[-startoption <start_option>] [-node
<node_name>]
    srvctl start mgmtlsnr [-node <node_name>]
    srvctl stop mgmtdb [-stopoption <stop_option>] [-force]
    srvctl stop mgmtlsnr [-node <node_name>] [-force]
```

关闭 MGMT 数据库，代码如下：

```
[grid@rac1 ~]$ srvctl stop mgmtdb -stopoption immediate -force
[grid@rac1 ~]$ srvctl status mgmtdb
Database is enabled
Database is not running.
```

启动 MGMT 数据库，代码如下：

```
[grid@rac1 ~]$ srvctl start mgmtdb
```

在启动时也可以指定节点，代码如下：

```
srvctl start mgmtdb -startoption open -n rac1
[grid@rac1 ~]$ srvctl status mgmtdb
Database is enabled
Instance -MGMTDB is running on node rac2
```

16.3.2 查看 MIGR 的资源

查看 MIGR 的资源，代码如下：

```
[grid@rac1 ~]$ crsctl stat  res -t
--------------------------------------------------------------------------------
Name           Target  State        Server                   State
details
--------------------------------------------------------------------------------
Local Resources
--------------------------------------------------------------------------------
…
ora.MGMT.GHCHKPT.advm
               OFFLINE OFFLINE      rac1                     STABLE
               OFFLINE OFFLINE      rac2                     STABLE
ora.MGMT.dg
               ONLINE  ONLINE       rac1                     STABLE
               ONLINE  ONLINE       rac2                     STABLE
…
ora.mgmt.ghchkpt.acfs
               OFFLINE OFFLINE      rac1                     STABLE
               OFFLINE OFFLINE      rac2                     STABLE
…
--------------------------------------------------------------------------------
Cluster Resources
--------------------------------------------------------------------------------
ora.LISTENER_SCAN1.lsnr
      1        ONLINE  ONLINE       rac1                     STABLE
ora.MGMTLSNR
      1        ONLINE  ONLINE       rac2
169.254.11.39 192.16
                                                             8.56.101,STABLE
…
ora.mgmtdb
      1        ONLINE  ONLINE       rac2                     Open,STABLE
…
```

16.3.3 查看 MGMT 数据库的告警日志和 Trace 文件

一般情况下，不需要查看 MGMT 数据库的 Trace 文件，如果要查看，告警日志和 Trace 文件在目录$GRID_BASE/rdbms/_mgmtdb/-MGMTDB/trace 中，代码如下：

```
[grid@rac1 trace]$ pwd
/u01/app/grid/diag/rdbms/_mgmtdb/-MGMTDB/trace
[grid@rac1 trace]$ ls
alert_-MGMTDB.log              -MGMTDB_j000_29801.trm
-MGMTDB_mz00_16944.trc         -MGMTDB_ora_6709.trm
drc-MGMTDB.log                 -MGMTDB_j000_436.trc
-MGMTDB_mz00_16944.trm         -MGMTDB_ora_6711.trc
-MGMTDB_aqpc_15413.trc         -MGMTDB_j000_436.trm
-MGMTDB_mz00_17576.trc         -MGMTDB_ora_6711.trm
-MGMTDB_aqpc_15413.trm         -MGMTDB_j000_6817.trc
-MGMTDB_mz00_17576.trm         -MGMTDB_ora_6713.trc
-MGMTDB_aqpc_15890.trc         -MGMTDB_j000_6817.trm
-MGMTDB_mz00_17868.trc         -MGMTDB_ora_6713.trm
-MGMTDB_aqpc_15890.trm         -MGMTDB_j000_6878.trc
-MGMTDB_mz00_17868.trm         -MGMTDB_ora_6714.trc
-MGMTDB_aqpc_4044.trc          -MGMTDB_j000_6878.trm
-MGMTDB_mz00_18332.trc         -MGMTDB_ora_6714.trm
```

第 17 章
数据库中的 CHM

CHM（Cluster Health Monitor，集群健康监视器）通过 OS API 来收集操作系统的统计信息，如内存、Swap 空间使用率、进程、I/O 使用率、网络等相关的数据。CHM 的信息收集是实时的，在 Oracle 11.2.0.3 之前，每秒收集一次数据；从 Oracle 11.2.0.3 开始，每 5 秒收集一次数据，并保存在 CHM 仓库中。这个收集时间间隔不能手动修改。

使用 CHM 的目的是在出现问题时，提供一个分析的依据，比如节点重启、挂起、实例被删除、性能下降等，这些问题都可以通过 CHM 收集的数据进行分析。通过对这些常量的监控，也可以提前知道系统的运行状态，以及资源是否异常。

在 Oracle 11.2.0.2 中，Oracle 把 CHM 整合到 GI 中，所以在 Oracle 11.2.0.2 的 Linux 和 Solaris 两个平台中，不需要单独安装 CHM。

从 CHM 的功能看，它与 OSWatcher 的功能非常像，但它们之间有以下一些区别。

- 收集频率不同：OSWatcher 的收集频率可以修改，而 CHM 的收集频率无法修改。在 Oracle 11.2.0.3 之前，CHM 每秒收集一次数据；从 Oracle 11.2.0.3 开始，CHM 每 5 秒收集一次数据。
- 占用系统资源不同：CHM 直接调用 OS 的 API 来减少开销，而 OSWatcher 直接调用 OS 命令，CHM 占用更少（小于 3%）的 CPU 资源。
- 收集信息不同：CHM 不会收集 top、traceroute 和 netstat 命令返回的数据。CHM 和 OSWatcher 可配合使用，如果资源限制只能使用一个方法，那么推荐使用 CHM。

- 运行优先级不同：OSWatcher 是运行在用户优先级上的，所以当 CPU 负载很高时是不能工作的，也就是说，CHM 可以收集 OSWatcher 收集不到的数据。

17.1　CHM 所需的磁盘大小

默认情况下，CHM 监控所有节点的数据需要 1GB 的空间，每个节点每天产生约 500MB 的数据。CHM 仓库默认将数据保留 3 天，所以 CHM 仓库的空间也在不断增加。

可以通过下面的公式来估算 CHM 所需的磁盘大小：

磁盘大小= 节点数×每个节点产生的数据大小×保留天数

例如，有 4 个节点，每个节点大概有 500MB 数据，保留 3 天数据，则 CHM 所需的磁盘大小是：

4×500×3=6000MB

CHM 的数据也是保存在 MGMT 数据库中的，如果 GIMR 对应的磁盘组空间不是很大，就会给数据存储带来一些压力，我们可以设置 CHM 数据的保留时间来减轻存储的压力。Oracle 建议保留 3 天的数据（最多只能保留 3 天）。

- 查看当前保留时间，代码如下：

```
[grid@RAC1 ~]$ oclumon manage -get repsize
CHM Repository Size = 136320 seconds
```

- 下面的代码提示想要保留更长的时间，必须先增加 CHM 仓库的大小。

```
[grid@RAC1 ~]$ oclumon manage -repos checkretentiontime 259200
The Cluster Health Monitor repository is too small for the desired
retention. Please first resize the repository to 3896 MB
```

- 修改 CHM 仓库的大小。

在 Oracle 18c 中，CHM 仓库的大小不能小于 2048MB（2GB），从下面的提示信息来看，已经将保留时间设置为 259 260s。

```
[grid@RAC1 ~]$ oclumon manage -repos changerepossize 3896
The Cluster Health Monitor repository was successfully resized.The
new retention is 259260 seconds.
```

注意，在 Oracle 11.2.0.2 中，默认 CHM 仓库的大小为 1GB；在 Oracle 18c

中，默认 CHM 仓库的大小为 2GB。

17.2 分析 CHM 数据

可以通过 diagcollection.pl 脚本或 oclumon 命令将存储在数据库中的 CHM 数据格式化输出显示。查看所有数据可能需要很长时间，可以根据特定时间段来查看数据。由于要调用操作系统命令来收集数据，所以使用 root 账户。

使用 diagcollection.pl 脚本分析数据，需要在主节点上运行该脚本。

17.2.1 分析所有之前收集的数据

在分析所有数据时，速度会很慢，除非有特殊需要，否则不建议一次性分析所有数据。

- 确认主节点，代码如下：

```
[root@RAC1 ~]$/u01/app/18.3.0/grid/bin/oclumon manage
-get master
Master = rac2
```

- 分析所有数据，代码如下：

```
[root@RAC1 ~]$/u01/app/18.3.0/grid/bin/
diagcollection.pl --collect -chmos
```

17.2.2 分析特定时间段的数据

1. 使用 diagcollection.pl 脚本

用 root 账户执行以下命令来分析 2018 年 7 月 30 号凌晨 1 点到 5 点的数据，然后在当前目录下生成一个.tar.gz 的压缩文件和一个.txt 文本文件。

（1）确认主节点，代码如下：

```
[root@RAC1 ~]$/u01/app/18.3.0/grid/bin/oclumon manage -get
master
Master = rac2
```

（2）分析指定时间段的数据，在主节点上（rac2）用 root 账户运行以下分析命令：

```
[root@RAC2 chm]$/u01/app/18.3.0/grid/bin/diagcollection.pl
```

```
--collect --crshome $ORA_CRS_HOME --chmos --incidenttime
07/30/201801:00:00 --incidentduration 05:00
    Production Copyright 2004, 2010, Oracle.  All rights reserved
    Cluster Ready Services （CRS） diagnostic collection tool
    ORACLE_BASE is /u01/app/grid
    Collecting Cluster Health Monitor （OS） data
    Version: 18.3.0.0.0
    Collecting OS logs
    Collecting sysconfig data
```

注意，这里的“07/30/201801:00:00”时间格式是固定的，必须是这个格式，而且年和小时之间没有空格。

（3）收集的文件如下：

```
    [root@rac2 chm]# ll
    total 23096
    -rw-r--r-- 1 root root 23443142 Jul 31 10:08
chmosData_rac2_20180731_1006.tar.gz
    -rw-r--r-- 1 root root   179098 Jul 31 10:08
osData_rac2_20180731_1006.tar.gz
    -rw-r--r-- 1 root root    23265 Jul 31 10:08
sysconfig_rac2_20180731_1006.txt
```

文件内容分别是：

- 详尽的资源使用信息，包括数据库进程信息、CPU、内存使用信息；
- 操作系统的日志（/var/log/message）；
- 操作系统的配置信息，包括 CPU、内存、磁盘、网络等。

2. 使用 oclumon 命令

如果使用 diagcollection.pl 脚本执行失败，则可以使用 oclumon 命令分析数据。根据过去 2 个小时收集的数据生成一个报告，代码如下：

```
    [root@RAC1 ~]$/u01/app/18.3.0/grid/bin/oclumon dumpnodeview
-allnodes -v -last "01:59:59" > /tmp/chm.log
```

也可以生成指定时间段的报告，代码如下：

```
    [root@RAC1 ~]$/u01/app/18.3.0/grid/bin/oclumon dumpnodeview
-allnodes -v -s "2018-07-29 22:15:00" -e "2018-07-30 03:15:00" >
/tmp/chm.log
```

17.2.3 CHMOSG 工具

可以使用图形化的管理工具 CHMOSG 来分析数据，该工具默认没有被安装，需要单独从 OTN 上下载。CHMOSG 工具会以图形化的方式详细展示相关数据。

CHMOSG 工具的下载网址：www.oracle.com/technetwork/products/clusterware/downloads/chm-os-ui-1554902.zip。

CHMOSG 工具的参考手册网址：http://www.oracle.com/technetwork/products/clusterware/downloads/chmosg-userman-1554904.pdf。

1. CHMOSG 的安装

CHMOSG 的安装步骤如下。

（1）解压缩下载好的压缩文件，代码如下：

```
[root@RAC1 chmosg]# ll
total 1052
-rw-r--r-- 1 root root 1075018 Jul 31 10:30 chm-os-ui-1554902.zip
[root@RAC1 chmosg]# unzip chm-os-ui-1554902.zip
[root@RAC1 chmosg]# ll
total 1056
drwxrwxr-x 2 root root     45 Mar 12  2012 bin
-rw-r--r-- 1 root root 1075018 Jul 31 10:30 chm-os-ui-1554902.zip
drwxrwxr-x 3 root root     44 Mar 12  2012 doc
drwxrwxr-x 2 root root     42 Mar 12  2012 install
drwxrwxr-x 2 root root     21 Mar 12  2012 jlib
-r--r--r-- 1 root root   3395 Mar 12  2012 README.txt
```

（2）在设置 CRFPERLBIN 环境变量之前，先设置 PERL 环境变量，并且要求 PERL 的版本要大于或等于 5.6.0。

查看 PERL 版本，代码如下：

```
[root@RAC1 install]# perl -v
This is perl 5, version 16, subversion 3 (v5.16.3) built for
x86_64-linux-thread-multi
(with 33 registered patches, see perl -V for more detail)
…
```

设置 PERL 环境变量，PERL 命令所在的目录默认是/usr/bin，代码如下：

```
[root@RAC1 install]# export CRFPERLBIN=/usr/bin
```

（3）设置 Java 环境变量，要求是 Java 1.5 及以上版本。

查看 Java 版本，代码如下：

```
[root@RAC1 install]# java -version
openjdk version "1.8.0_161"
OpenJDK Runtime Environment (build 1.8.0_161-b14)
OpenJDK 64-Bit Server VM (build 25.161-b14, mixed mode)
```

设置 Java Home 环境变量，代码如下：

```
[root@RAC1 ~]# export
CRF_JAVA_HOME=/usr/lib/jvm/java-1.8.0-openjdk-1.8.0.161-2.b14.el7.
x86_64/jre/
```

（4）创建 CHMOSG 安装目录，保证执行 CHMOSG 命令的用户有对该目录的读写权限，代码如下：

```
[root@RAC1 ~]# mkdir -p /usr/lib/oracf/chmosg
```

（5）安装 CHMOSG，代码如下：

```
[root@RAC1 install]# ./chminstall -i /usr/lib/oracf/chmosg
Can't locate Env.pm in @INC (@INC contains: /usr/local/lib64/perl5
/usr/local/share/perl5 /usr/lib64/perl5/vendor_perl
/usr/share/perl5/vendor_perl /usr/lib64/perl5 /usr/share/perl5 .)
at ./crfinst.pl line 30.
BEGIN failed--compilation aborted at ./crfinst.pl line 30.
```

报错原因是没有安装 perl-Env 包，解决方法是安装 perl-Env 包，命令如下：

yum install perl-Env -y

再次安装 CHMOSG，代码如下：

```
[root@RAC1 install]# ./chminstall -i /usr/lib/oracf/chmosg
ORACLE_HOME found. Using Java at /u01/app/18.3.0/grid/jdk
Installation completed sucessfully at /usr/lib/oracf/chmosg...
```

（6）运行 CHMOSG，代码如下：

```
[root@RAC1 bin]# cd /usr/lib/oracf/chmosg/bin/
[root@RAC1 bin]# ./chmosg
```

运行后即可通过图形化界面实时查看与操作系统相关的信息，如 CPU、内存、网络、磁盘 I/O 等，如图 17-1 所示。

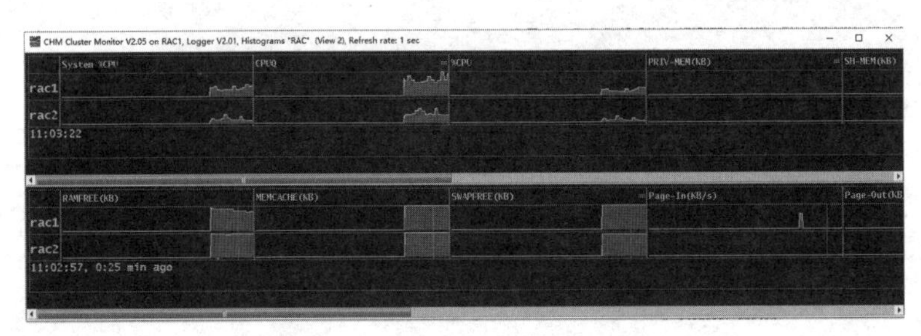

图 17-1 实时查看与操作系统相关的信息

2. CHMOSG 的使用

查看帮助信息，代码如下：

```
[root@RAC1 bin]# ./chmosg -h
Java HotSpot (TM) 64-Bit Server VM warning: Using incremental CMS
is deprecated and will likely be removed in a future release
Cluster Health Monitor V2.05
Usage: crfgui [flags] [params]
  -h        print this help
  -?        print this help
…
```

连接指定节点，代码如下：

```
[root@RAC1 bin]# ./chmosg -m rac1
```

双击一个节点即可查看该节点的信息，如图 17-2 所示。

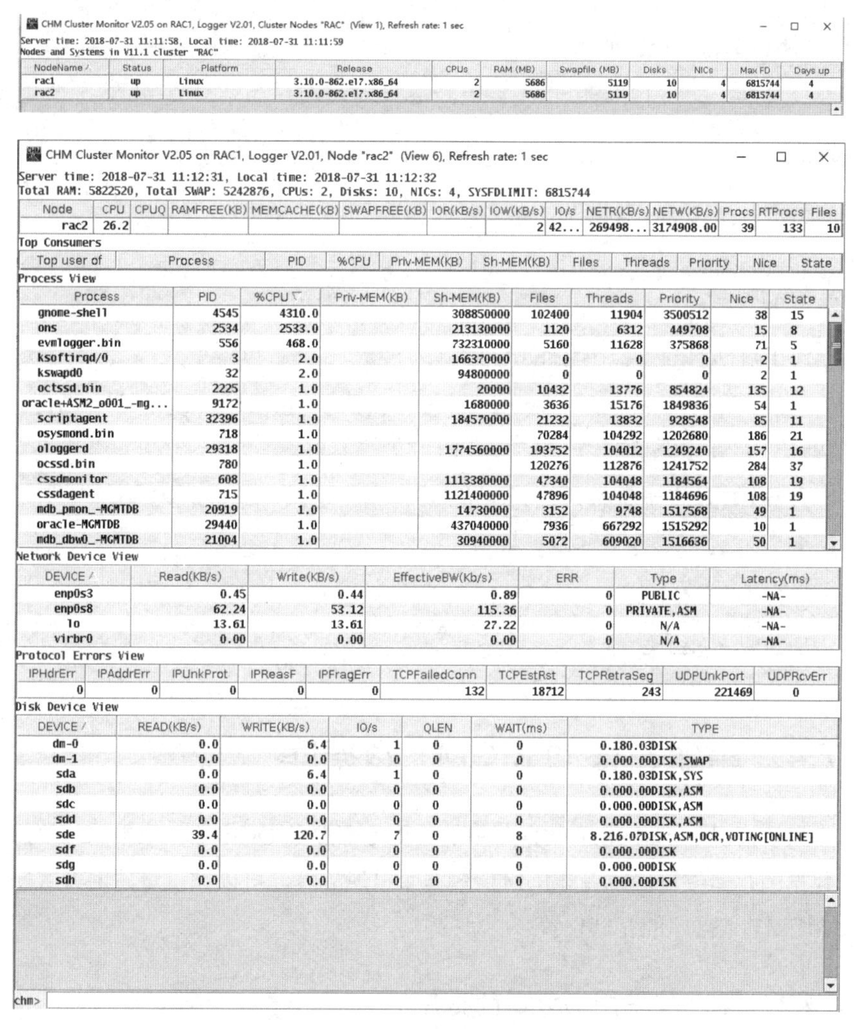

图 17-2　查看节点的详细信息

17.3　管理 CHM

在 Oracle 11.2 及以上版本中，CHM 可以通过 ora.crf 资源来管理。

关闭 CHM，代码如下：

```
[grid@RAC1 ~]$ crsctl stop res ora.crf -init
CRS-2673: Attempting to stop 'ora.crf' on 'rac1'
CRS-2677: Stop of 'ora.crf' on 'rac1' succeeded
```

启动 CHM，代码如下：

```
[grid@RAC1 ~]$ crsctl start res ora.crf -init
CRS-2672: Attempting to start 'ora.crf' on 'rac1'
CRS-2676: Start of 'ora.crf' on 'rac1' succeeded
```

禁用自动启动，代码如下：

```
[root@RAC1 software]# /u01/app/18.3.0/grid/bin/crsctl modify resource ora.crf -attr "AUTO_START=never" -init
```

启用自动启动，代码如下：

```
[root@RAC1 software]# /u01/app/18.3.0/grid/bin/crsctl modify resource ora.crf -attr "AUTO_START=always" -init
```

查看 CHM 资源状态，代码如下：

```
[grid@RAC1 ~]$ crsctl stat res ora.crf -init
NAME=ora.crf
TYPE=ora.crf.type
TARGET=ONLINE
STATE=ONLINE on rac1
```

17.4 重建、移动 MGMT 数据库

默认情况下，MGMT 数据库中的数据是存放在 OCR 磁盘组中的，要想移动数据的存储位置，或者 MGMT 数据库出现故障，可以通过重建的方式来修改 MGMT 数据库的存储位置。

17.4.1 停止并禁用 ora.crf 资源

因为 CHM 的数据是存放在 MGMT 数据库中的，所以在重建之前需要停止使用 CHM 的 ora.crf 资源。

在所有节点使用 root 账户执行如下命令：

```
[root@rac1 ~]# crsctl stop res ora.crf -init
CRS-2673: Attempting to stop 'ora.crf' on 'rac1'
CRS-2677: Stop of 'ora.crf' on 'rac1' succeeded

[root@rac1 ~]# crsctl modify res ora.crf -attr ENABLED=0 -init
[root@rac1 ~]#
```

```
[root@rac2 ~]# crsctl stop res ora.crf -init
CRS-2673: Attempting to stop 'ora.crf' on 'rac2'
CRS-2677: Stop of 'ora.crf' on 'rac2' succeeded

[root@rac2 ~]# crsctl modify res ora.crf -attr ENABLED=0 -init
[root@rac2 ~]#
```

注意，这里的 ora.mgmtlsnr 和 ora.mgmtdb 资源不能停止，否则执行 DBCA 命令时会报错。

17.4.2　使用 DBCA 命令删除 MGMT 数据库

查看 MGMT 数据库的运行节点，代码如下：

```
[root@rac2 ~]#srvctl status mgmtdb
Database is enabled
Instance -MGMTDB is running on node rac2
```

上面的代码显示 MGMT 数据库在节点 2 上运行，所以在节点 2 上用 grid 账户执行 DBCA 命令删除 MGMT 数据库，代码如下：

```
[grid@rac2 ~]$ dbca -silent -deleteDatabase -sourceDB -MGMTDB
[WARNING] [DBT-19202] The Database Configuration Assistant will
delete the Oracle instances and datafiles for your database. All
information in the database will be destroyed.
Prepare for db operation
32% complete
Connecting to database
35% complete
39% complete
42% complete
45% complete
48% complete
52% complete
65% complete
Updating network configuration files
68% complete
Deleting instance and datafiles
84% complete
100% complete
Database deletion completed.
```

```
Look at the log file
"/u01/app/grid/cfgtoollogs/dbca/_mgmtdb/_mgmtdb.log" for further
details.
[grid@rac2 ~]$
```

17.4.3 重建 MGMT 数据库的 CDB

在 RAC 集群的任意节点用 grid 账户执行如下命令：

```
$ <GI_HOME>/bin/dbca -silent -createDatabase -sid -MGMTDB
-createAsContainerDatabase true -templateName MGMTSeed_Database.dbc
-gdbName _mgmtdb -storageType ASM -diskGroupName <+NEW_DG>
-datafileJarLocation $GRID_HOME/assistants/dbca/templates
-characterset AL32UTF8 -autoGeneratePasswords -skipUserTemplateCheck
```

结果如下：

```
[grid@rac2 ~]$ dbca -silent -createDatabase -sid -MGMTDB
-createAsContainerDatabase true -templateName MGMTSeed_Database.dbc
-gdbName _mgmtdb -storageType ASM -diskGroupName +MGMT
-datafileJarLocation $ORACLE_HOME/assistants/dbca/templates
-characterset AL32UTF8 -autoGeneratePasswords -skipUserTemplateCheck
Prepare for db operation
10% complete
Registering database with Oracle Grid Infrastructure
14% complete
Copying database files
43% complete
Creating and starting Oracle instance
45% complete
49% complete
54% complete
58% complete
62% complete
Completing Database Creation
66% complete
69% complete
71% complete
Executing Post Configuration Actions
100% complete
Database creation complete. For details check the logfiles at:
 /u01/app/grid/cfgtoollogs/dbca/_mgmtdb.
Database Information:
Global Database Name:_mgmtdb
```

```
    System Identifier (SID) :-MGMTDB
    Look at the log file "/u01/app/grid/cfgtoollogs/dbca/_mgmtdb
/_mgmtdb0.log" for further details.
    [grid@rac2 ~]$

    [grid@rac2 ~]$ srvctl status mgmtdb
    Database is enabled
    Instance -MGMTDB is running on node rac2
```

17.4.4 使用 DBCA 命令创建 PDB

在任意节点上，用 grid 账户执行 DBCA 命令创建 PDB，代码如下：

```
    $ <GI_HOME>/bin/dbca -silent -createPluggableDatabase -sourceDB
-MGMTDB -pdbName <CLUSTER_NAME> -createPDBFrom RMANBACKUP
-PDBBackUpfile <GI_HOME>/assistants/dbca/templates/pdbseed.dfb
-PDBMetadataFile <GI_HOME>/assistants/dbca/templates/pdbseed.xml
-createAsClone true -internalSkipGIHomeCheck
```

查询集群的名称，代码如下：

```
    [grid@rac2 ~]$ cemutlo -n
    rac
    [grid@rac2 ~]$
```

笔者的集群名是 rac，如果集群名中间有“-”，则必须替换成“_”，如 rac-scan 需要替换成 rac_scan。这里直接执行命令，结果如下：

```
    [grid@rac2 templates]$ dbca -silent -createPluggableDatabase
-sourceDB -MGMTDB -pdbName rac -createPDBFrom RMANBACKUP
-PDBBackUpfile $ORACLE_HOME/assistants/dbca/templates/pdbseed.dfb
-PDBMetadataFile $ORACLE_HOME/assistants/dbca/templates/pdbseed.xml
-createAsClone true -internalSkipGIHomeCheck
    Prepare for db operation
    13% complete
    Creating Pluggable Database
    15% complete
    19% complete
    23% complete
    31% complete
    39% complete
    53% complete
    Completing Pluggable Database Creation
    60% complete
```

```
Executing Post Configuration Actions
100% complete
Pluggable database "rac" plugged successfully.
Look at the log file
"/u01/app/grid/cfgtoollogs/dbca/_mgmtdb/rac/_mgmtdb.log" for
further details.
```

17.4.5　验证 MGMT 数据库

验证 MGMT 数据库，代码如下：

```
[grid@rac2 templates]$ export ORACLE_SID=-MGMTDB
[grid@rac2 templates]$ sqlplus / as sysdba
SQL*Plus: Release 18.0.0.0.0 - Production on Thu Sep 13 20:43:44
2018
Version 18.3.0.0.0
Copyright (c) 1982, 2018, Oracle.  All rights reserved.
Connected to:
Oracle Database 18c Enterprise Edition Release 18.0.0.0.0 -
Production
Version 18.3.0.0.0
SQL> show pdbs
   CON_ID CON_NAME                OPEN MODE  RESTRICTED
---------- ------------------------------ ---------- ----------
     2 PDB$SEED              READ ONLY  NO
     3 RAC                   READ WRITE NO
SQL> select file_name from dba_data_files union select member
file_name from V$logfile;
FILE_NAME
------------------------------------------------------------------
------------------
+MGMT/_MGMTDB/DATAFILE/sysaux.261.986759947
+MGMT/_MGMTDB/DATAFILE/system.270.986759927
+MGMT/_MGMTDB/DATAFILE/undotbs1.269.986759963
+MGMT/_MGMTDB/ONLINELOG/group_1.263.986760019
+MGMT/_MGMTDB/ONLINELOG/group_2.262.986760019
+MGMT/_MGMTDB/ONLINELOG/group_3.264.986760019
6 rows selected.
```

17.4.6　启动 ora.crf 资源

在所有节点上，用 root 账户执行以下命令：

```
[root@rac1 ~]# crsctl modify res ora.crf -attr ENABLED=1 -init
[root@rac1 ~]# crsctl start res ora.crf -init
CRS-2672: Attempting to start 'ora.crf' on 'rac1'
CRS-2676: Start of 'ora.crf' on 'rac1' succeeded
[root@rac1 ~]# oclumon manage -get master

Master = rac2
[root@rac1 ~]#
```

附录 A

Oracle 软件版本和生命周期

1. Oracle 版本号变化说明

Oracle 版本号在 2018 年发生了一些非常明显的变化，最主要的表现就是从 12c 直接跳到了 18c，这让很多人都不习惯。

在 Oracle 12c 之前，Oracle 版本号的命名规则是一样的，我们可以从 Oracle 官方文档上找到出处，如图 A-1 所示是官方文档中 Oracle 11g 版本号的命名规则。

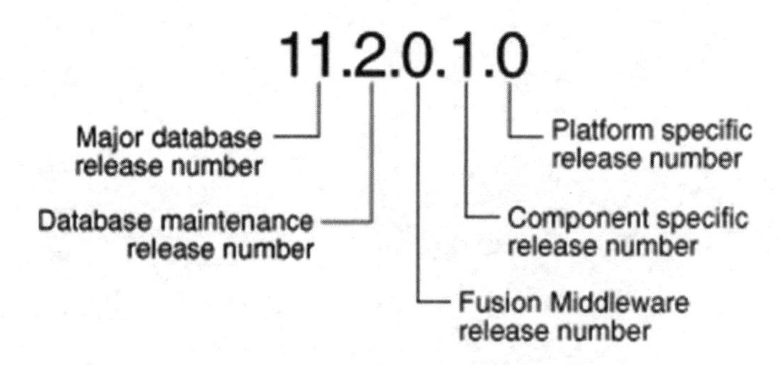

图 A-1　Oracle 11g 版本号的命名规则

第一位数字是数据库的主版本号，通常也标志一些新功能，比如 10g、11g、12c。

第二位数字是数据库维护版本号，代表一个维护发行版级别，也可能包含一些新的特性，比如 11gR1（11.1）、11gR2（11.2）。

第三位数字是融合中间件版本号，这个数字和 Oracle 数据库版本没有关系。

第四位数字是组件的特定版本号，与 PSR 对应。比如，11.2.0.1 是一个主版本，那么 11.2.0.2、11.2.0.3 就是两个不同的补丁集。

第五位数字是平台特定版本号，与 PSU 对应，虽然在描述 PSU 时会用到数据库版本的第 5 位，比如 Database PSU 11.2.0.3.5，但实际上安装 PSU 后并不会真正改变数据库的版本号，从 v$version 视图中看到的版本号还是 4 位（11.2.0.3.0），第 5 位数字仍然是 0。

下面看一下官方文档中 Oracle 12c 版本号的命名规则，如图 A-2 所示。

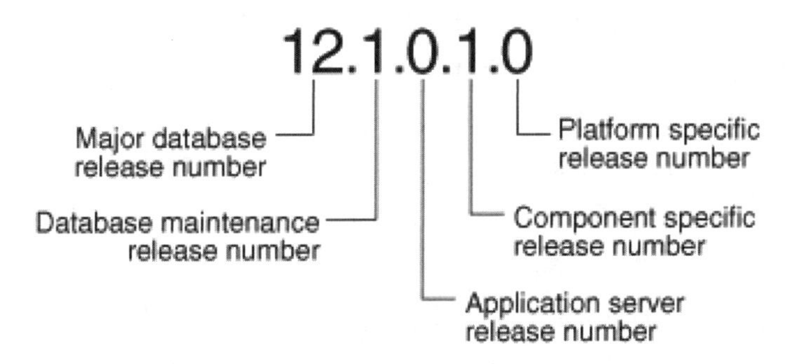

图 A-2　Oracle 12c 版本号的命名规则

从图 A-2 中可以看出，除第三位数字的解释发生变化外，其余数字解释和 Oracle 11g 版本号的数字解释并没有区别。到了 Oracle 18c，虽然软件的版本号还是由 5 位数字来表示，但这 5 位数字的意义发生了明显的变化，可以查看官方文档中的版本号命名规则，如图 A-3 所示。

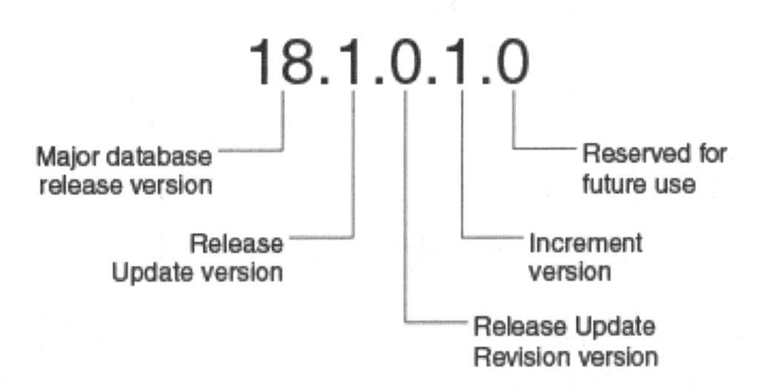

图 A-3　Oracle 18c 版本号的命名规则

从 Oracle 18c 开始，Oracle 的软件版本号由发行版（Version）和完全发行版（Version_full）组成。

发行版的格式是：主版本号. 0.0.0.0。主版本号由 Oracle 软件的发布年度来指定，比如，如果 Oracle 软件是 2018 年被发布的，那么主版本号就是 18。

完全发行版的格式根据软件发布的年度，以及 RU 和 RUR 发布的季度进行指定。也就是说，从 2018 年开始，Oracle 数据库使用了一个新的命名格式：Year.Update.Revision。比如 18.1.0，这里分别表示数据库软件发布的年度、PU 发布的季度和 RUR 发布的季度。

从图 A-3 来看，版本号还是五位数字，下面分别介绍它们所代表的含义。

第一位数字是 Oracle 数据库的主版本号，比如 Oracle 18c、Oracle 12c。从 Oracle 18c 开始，第一个数字表示 Oracle 数据库版本发布的最初年度，比如 2018 年是 Oracle 18c（18.0.0.0.0）的最初发布年度。

第二位数字是 Oracle RU（Release Update）的发布季度，比如 18.3。

第三位数字是 Oracle RUR（Release Updates Revision）的发布季度，比如 18.1.1、18.2.1、18.3.0。

第四位数字是数据库的增量版本，这种命名方法可以用于未来版本的更新，并且可以应用在 18c 之前的所有版本上，比如 12.1.0.1、12.2.0.1。

第五位数字是为将来使用而预留的。

这里的 RU 和 RUR 是 Oracle 12cR2 以后补丁体系的一个变化。如图 A-4 所示是官方提供的 Oracle 18c 以后的版本命名变化。

	Production	Q2	Q3	Q4	Q1	Q2	Q3	Q4
Update	18.1.0	18.2.0	18.3.0	18.4.0	18.5.0	18.6.0	18.7.0	18.8.0
Revision #1			18.2.1	18.3.1	18.4.1	18.5.1	18.6.1	18.7.1
Revision #2				18.2.2	18.3.2	18.4.2	18.5.2	18.6.2
Update					19.1.0	19.2.0	19.3.0	19.4.0
Revision#1							19.2.1	19.3.1
Revision #2								19.2.2

图 A-4 Oracle 18c 以后的版本命名变化

简单地说，就是从 Oracle 18c 开始，数据库软件的版本号真正有意义的就是前三位，后面两位都是预留的。

2. Oracle 软件的生命周期说明

Oracle 软件的生命周期分为以下三种类型。

- Premier Support（5 年）：这种类型给 Oracle 数据库、Oracle Fusion 中间件和 Oracle 应用程序提供全面的维护和软件升级，一般从软件发布开始算起。
- Extended Support（3 年）：这种类型为 Oracle 数据库、Oracle Fusion 中间件和 Oracle 应用程序提供额外的维护和升级，但会产生额外的费用。
- Sustaining Support（无限期）：只要还在使用 Oracle 软件，就可以提供维护服务，但支持有限。

以上三种类型的具体区别可以参考图 A-5，这里直接从官网截图，不再进行翻译。

	Premier Support	Extended Support	Sustaining Support
Major product and technology releases	✔	✔	✔
24x7 assistance with service requests	✔	✔	✔
Access to My Oracle Support including Knowledge Base	✔	✔	✔
Software updates	✔	✔	Pre-existing
Security alerts and updates	✔	✔	Pre-existing
Critical patch updates	✔	✔	Pre-existing
Tax, legal, and regulatory updates	✔	✔	Pre-existing
Upgrade tools and scripts	✔	✔	Pre-existing
Access to Platinum Services	✔	✔	✘
Certification with most existing Oracle products/versions	✔	✔	Pre-existing
Certification with most existing third party products	✔	✔	Pre-existing
Certification with most new third party products	✔	✘	✘

图 A-5　软件生命周期三种类型的具体区别

在 Oracle 官网（http://www.oracle.com/us/support/library/ lifetime-support-technology- 069183.pdf）上，Oracle 不同版本的不同生命周期的时间如图 A-6 所示。

Release	GA Date	Premier Support Ends	Extended Support Ends	Sustaining Support Ends
8.1.7	Sep 2000	Dec 2004	Dec 2006	Indefinite
9.2	Jul 2002	Jul 2007	Jul 2010	Indefinite
10.1	Jan 2004	Jan 2009	Jan 2012	Indefinite
10.2	Jul 2005	Jul 2010	Jul 2013	Indefinite
11.1	Aug 2007	Aug 2012	Aug 2015	Indefinite
11.2	Sep 2009	Jan 2015	Dec 2020	Indefinite
Enterprise Edition 12.1	Jun 2013	Jul 2018	Jul 2021	Indefinite
Standard Edition (SE) 12.1	Jun 2013	Aug 2016	Not Available	Indefinite
Standard Edition One (SE1) 12.1	Jun 2013	Aug 2016	Not Available	Indefinite
Standard Edition 2 (SE2) 12.1	Sep 2015	Jul 2018	Jul 2021	Indefinite
12.2	Mar 2017	Mar 2022	Mar 2025	Indefinite

图 A-6 Oracle 不同版本的不同生命周期的时间

在 MOS 文档（ID 742060.1）中可以看到 Oracle 软件版本号发展的路线图，如图 A-7 所示。

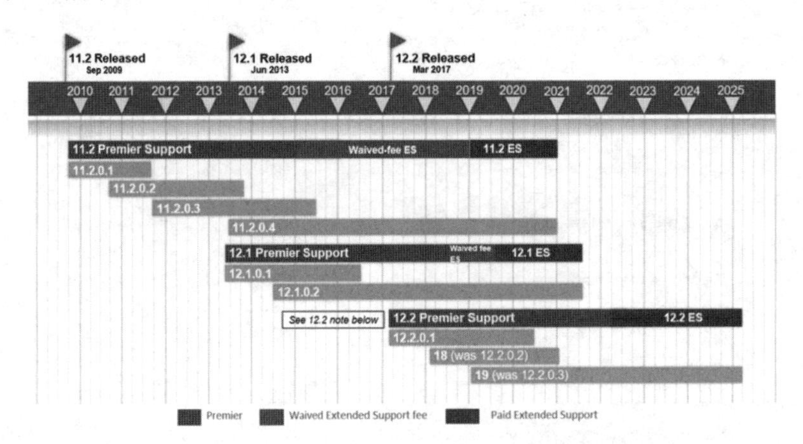

图 A-7 Oracle 软件版本号发展的路线图

根据前面的说明，Oracle 版本号从 18c 开始真正有意义的是前 3 位数字，在 Oracle 18c 和 Oracle 19c 版本号中也列出了与之前的版本的对应关系，比如 18c 等于 12.2.0.2、19c 等于 12.2.0.3。这个应该是一个过渡阶段，待 12c 的生命周期结束以后，在软件版本号发展的路线图上应该就不会看到这种情况了。

3. Oracle 补丁说明

与 Oracle 数据库版本号一样，Oracle 补丁的相关体系结构也一直在变化，尤其是 Oracle 12c 之后的版本。下面介绍一下这种变化，防止在多种信息同时出现时，读者感到混乱。这里只做一些概述，想了解补丁详细说明的读者，可以参考 Oracle MOS 文档（文档 ID：2285040.1）。

（1）Oracle 12.2.0.1 之前的补丁说明

在 Oracle 12.2.0.1 版本之前，Oracle 数据库提供了两种类型的补丁：Proactive 和 Reactive。

Reactive 补丁：通常也被称为临时补丁（Interim Patch）或一次性补丁（One-off Patch），是对特定的缺陷、版本、平台组合而按需提供的补丁。这些修复通常在下一个相关的补丁集中被包含进来，每个补丁集都有一个“代码冻结”日期，在该日期后，只有阻碍运行的关键 bug 的修复才会被包含进来。如果一个修复错过了这次的补丁集，则将在后续的补丁集中（如果有的话）被包含进来。

Proactive 补丁：解决影响特定配置的有较大影响的 bug，包含已证实的低风险修复，在 Oracle MOS 上可以下载。

下面主要介绍 Proactive 补丁的几种类型。

- Security Patch Update（SPU）：作为 Oracle 的 Critical Patch Update（CPU）计划的一部分而发布的安全修复程序的集合，以预定义的季度计划交付，Oracle CPU 的发布日期大约在 1 月、4 月、7 月和 10 月的最接近 15 日的星期二。 SPU 正在从 Oracle 12c 版本中被淘汰，CPU 将在 Bundle Patch 或 PSU 中发布。
- Patch Set Update（PSU）：已证实有较大影响的 bug 的修复程序集合，包括作为 CPU 计划的一部分而发布的安全修复程序，在集合中不包含任何能改变应用程序行为的优化器更改或修复可能跨多个产品的组件。以预定义的季度计划交付，数据库 PSU 和 GI PSU 通常可以通过 RAC 滚动应用和备库首先安装，但 OJVM PSU 既不可进行 RAC 滚动安装，也不可在备库中首先安装。
- Bundle Patch（BP）：用于解决给定功能、产品或配置的 bug 的一组修复集合。BP 是 PSU 的一个超集，可以跨越多个产品组件。例如，

Database Patch for Exadata 是既包含数据库又包含 GI 的修复。BP 也以预定义的计划交付，可能比 PSU 发布频率更高，可以通过 RAC 滚动应用和备库首先安装。

- Quarterly Full Stack Download Patch / Combo Patch：若干不同的补丁打包在一起发布。例如，Quarterly Full Stack Download Patch for Exadata 把 Quarterly Database Patch for Exadata 与 OJVM PSU，还有其他的 Exadata 系统补丁打包在一个下载链接中。
- Other Proactive Patches：在正常的 SPU、PSU、BP 周期之外，Oracle 会针对特殊用途发布一些 Proactive 补丁。例如，针对客户系统使用最新时区数据的需求，每 6 个月发布一次特殊的 time-zone 补丁。这样的补丁通常作为临时补丁被发布。

各种 Proactive 补丁的关系如图 A-8 所示。

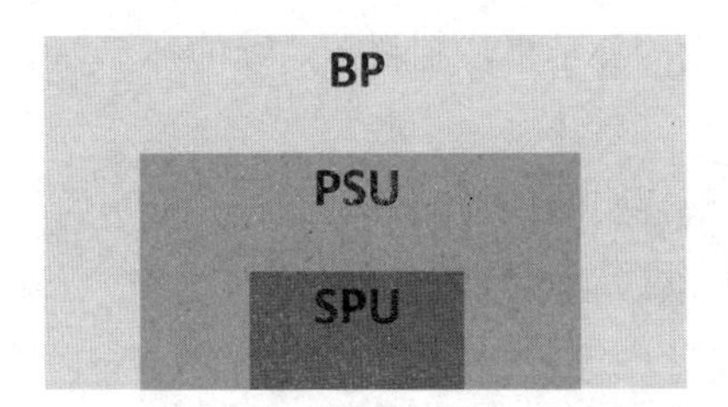

图 A-8 各种 Proactive 补丁的关系

- SPU 仅包含 CPU 计划的安全修复。
- PSU 包含 CPU 计划的安全修复及其他高影响、低风险的关键性 bug 的修复。
- BP 包含所有 PSU 修复及其他额外的修复。

在安装时只能使用 SPU、PSU、BP 三种补丁方式之一。所有方式都允许安装临时补丁，但是根据不同的补丁方式，临时补丁的版本也有可能不同。Windows 平台较特殊，因为 Windows 平台不支持一般的临时补丁。

（2）Oracle 12.2.0.1 之后的补丁说明

从 2017 年 7 月开始，Oracle 对数据库和 GI（Grid Infrastructure）12.2 及之后版本的主动修补程序进行了更改，也就是从 Oracle 18c（对应 Oracle 12.2.0.2）开始，数据库产品的新版本发布改为每年一次，并且不再发布补丁集。传统的 Patchset Update 和 Database Proactive Bundle Patch 在 GI 12.2 之后将不再发布。

传统术语 Patchset、Patchset Update 和 Database Bundle Patch 在 GI 12.2 中也不再适用。GI 12.2 及其之后的版本将采用新的 RU（Release Updates）和 RUR（Release Update Revisions）。

注意这里的更改只适用于数据库和 GI 12.2 及之后的版本，GI 12.1 之前的版本仍然使用传统的 PSU/BP 流程及版本编号系统。

为了支持与安全相关的修复及高优先级的非安全修复，Oracle 会在每年的 1 月、4 月、7 月和 10 月即每个季度各发布一个 RU。Oracle 每个季度发布的更新包含用户最有可能遇到的错误的修复。

- 查询优化器的错误修复，在之前版本的 PSU 和 BP 中没有包含的这些修复将被放到更新中，但默认是禁用的。
- 更新中包含与安全相关的补丁。
- 更新会经过广泛的测试，包括功能测试、压力测试、性能测试和破坏性测试。
- 及时应用更新可以减小遇到已知问题的可能性。
- 更新在 RAC 环境下可以使用滚动的方式不停机安装。

除每个季度发布的更新外，还会在每个季度发布 Release Update Revisions（Revisions），包含对更新的回退修复和最新的安全方面的修复。

在每个更新发布后的 6 个月内，会有两个针对这个更新的修订，如 Release.Update .1 和 Release.Update.2，这里的“1”“2”代表修订版本。

Oracle 推荐用户使用最新的更新版本，这样可以避免很多已知的问题，并且可以避免申请很多小补丁，可以显著地减少补丁维护操作。如果系统已达到稳定状态，并希望优先考虑安全更新而不是功能修复，则可以使用修订。当使用 Release.Update.1 时，落后更新的内容 3 个月；当使用 Release.Update.2 时，落后更新的内容 6 个月。通过选择延迟 3 个月或 6 个月更新内容来进行数据库软件的维护，仍有可能会碰到最新更新中包含的已知问题。

在更新和修订之间可以来回进行切换，但是是有限制的，新的补丁必须是之前补丁的超集。为了避免补丁冲突，用户应该坚持一贯的方式，即在每个季度维护周期中始终使用相同的修订级别，比如 Release.Update.0、Release.Update.1 或 Release.Update.2。

如图 A-9 所示是 Oracle 12.2.0.1 数据库版本 RU、RUR 的命名规则。

Production	July	October	Jan	April	July
12.2.0.1	Database Release Update 12.2.0.1.<build-date>	Database Release Update 12.2.0.1.<build-date>	Database Release Update 12.2.0.1.<build-date>	←	RU
		Database July 2017 Release Update Revision 12.2.0.1.<build-date>	Database Oct2017 Release Update Revision 12.2.0.1.<build-date>	←	RUR #1
			Database July 2017 Release Update Revision 12.2.0.1.<build-date>	←	RUR #2

图 A-9 Oracle 12.2.0.1 数据库版本 RU、RUR 的命名规则

- Release Update（RU）：数据库 Release Update 12.2.0.1.<build-date>。
- Release Update Revision（RUR）：数据库<Quarter> Release Update Revision 12.2.0.1.<build-date>。

这里的<Quarter>的格式是'MMM YYYY'，<build-date>的格式是 YYMMDD。

虽然 Oracle 将补丁进行了更改，但 RU 和 RUR 依旧使用现有的 Opatch 技术来安装 RU/RUR。

再回到我们之前提到的 18c 的版本号，18c 以后的数据库版本使用 3 位数字编码格式：年.更新.发布（Year.Update.Revision)，比如 18.1.0。和这里讲的内容对应起来，就容易理解为什么 Oracle 要这样来修改版本名称了。

（3）补丁版本号

不同的补丁方法用不同的方式来表示版本信息，如表 A-1 所示。

表 A-1 不同的补丁方法的版本信息

术语	版本号	举例
Major Release	版本的前 2 个字段指定（小数点分隔的字段）	11.2
Base Release	Major Release 版本的前 4 个字段	11.2.0.1
Patch Set Release (PSR)	版本的第 4 个字段指定	11.2.0.4
Patch Set Update (PSU)	版本的第 5 个字段指定	11.2.0.4.160419
Bundle Patch (BP)	版本的第 5 个字段指定，同时带有文本说明 Bundle 的系列	Exadata Database Bundle Patch 11.2.0.4.160419
Security Patch Update (SPU)	由月和年指定	11.2.0.4 Jan 2015 SPU

这里注意版本号的变化，从 2015 年 11 月开始，Oracle 数据库的 Bundle Patches、Patch Set Updates 和 Security Patch Updates 的版本号有了新的格式。如

图 A-10 和图 A-11 所示是从 Oracle 官网截的图。

Related Information

18.0.0.0

Description	Database Update	GI Update	Windows Bundle Patch
JUL2018 (18.3.0.0.180717)	28090523	28096386	NA
APR2018 (18.2.0.0.180417)	27676517	27681568	NA

图 A-10　Oracle 官网截图 1

Related Information

12.1.0.2

Description	PSU	GI PSU	Proactive Bundle Patch	Bundle Patch (Windows 32bit & 64bit)
JUL2018 (12.1.0.2.180717)	27547329	27967747	27968010	27937907
APR2018 (12.1.0.2.180417)	27338041	27468957	27486326	27440294
JAN2018 (12.1.0.2.180116)	26925311	27010872	27010930	27162953
OCT2017 (12.1.0.2.171017)	26713565	26635815	26635880	26720785
AUG2017(12.1.0.2.170814)	26609783	26610308	26610322	26161726
JUL2017 (12.1.0.2.170718)	25755742	25901062	26022196	26161724
APR2017 (12.1.0.2.170418)	25171037	25434003	25433352	25632533
JAN2017 (12.1.0.2.170117)	24732082	24917825	24968615	25115951

图 A-11　Oracle 官网截图 2

可以清楚地看到新的格式，用发布日期 YYMMDD 格式替代了 Bundle 版本的第 5 个数字。

- YY 是年份的最后 2 位数字。
- MM 是数字月份（2 位）。
- DD 是月份中的数字日期（2 位）。

从这个新版本格式可以直观地看到 bundle patch 来自哪个时间段，特别是哪些补丁来自相同的 Critical Patch Update 版本。

在前面也提到过，只有版本的前 4 个字段显示在数据库的 views/trace

banners 中，补丁信息不能在视图中查询。可以通过 Opatch 命令查看第 5 个字段的变化。使用 Opatch lsinventory 命令查询会显示 Patch Set 或 Base Release 及安装的补丁列表，该补丁列表通常包含一行或几行描述性语句来表示 SPU、PSU、BP 正在使用的补丁方法及版本。